SpringerBriefs in Applied Sciences and Technology

Mathematical Methods

Series Editors

Anna Marciniak-Czochra, Institute of Applied Mathematics, IWR, University of Heidelberg, Heidelberg, Germany

Thomas Reichelt, Emmy-Noether Research Group, Universität Heidelberg, Heidelberg, Germany

More information about this subseries at http://www.springer.com/series/11219

Anatoliy Malyarenko · Martin Ostoja-Starzewski ·
Amirhossein Amiri-Hezaveh

Random Fields of Piezoelectricity and Piezomagnetism

Correlation Structures

Springer

Anatoliy Malyarenko
Division of Mathematics and Physics
Mälardalen University
Västerås, Sweden

Amirhossein Amiri-Hezaveh
Department of Mechanical
Sciences and Engineering
University of Illinois
at Urbana-Champaign
Urbana, IL, USA

Martin Ostoja-Starzewski
Department of Mechanical Science
and Engineering
Institute for Condensed Matter Theory
Beckman Institute
University of Illinois
at Urbana-Champaign
Urbana, IL, USA

ISSN 2191-530X ISSN 2191-5318 (electronic)
SpringerBriefs in Applied Sciences and Technology
ISSN 2365-0826 ISSN 2365-0834 (electronic)
SpringerBriefs in Mathematical Methods
ISBN 978-3-030-60063-1 ISBN 978-3-030-60064-8 (eBook)
https://doi.org/10.1007/978-3-030-60064-8

Mathematics Subject Classification: 00A69, 74Axx, 60G60

This Springer imprint is published by the registered company Springer Nature Switzerland AG
The registered company address is: Gewerbestrasse 11, 6330 Cham, Switzerland

Preface

Spatially random materials pose mathematical challenges to theoretical physics and mechanics. These problems are exacerbated in the case of coupled field phenomena such as piezoelectricity and piezomagnetism, with the practical motivation coming from modern technology. The coupling of electrical/magnetic fields with elastic materials gives rise to rank 3 tensor-valued random fields, which have previously evaded description. The main purpose of this book is to provide an answer in terms of second-order wide-sense homogeneous and isotropic tensor-valued random fields. A complete description is given of such fields taking values in the three-dimensional linear space of piezoelectric tensors with at least D_2 symmetry.

Working from the standpoint of invariance of physical laws with respect to the choice of a coordinate system, spatial domain representations, as well as their spectra, are given in full detail for the orthotropic, tetragonal, and cubic crystal systems. Using group representation theory as the foundation, the derivations are done in terms of the complete description of the one- and two-point correlation tensors of the above class of fields as well as the spectral expansions of the fields in terms of stochastic integrals.

Västerås, Sweden/Urbana, USA
February 2020

Anatoliy Malyarenko
Martin Ostoja-Starzewski
Amirhossein Amiri-Hezaveh

Acknowledgements

The first named author is grateful to his colleagues at Mälardalen University for creating a friendly working and research environment.

The second and third named authors were partially supported by the NSF under grants CMMI-1462749 and IP-1362146 (I/UCRC on Novel High Voltage/Temperature Materials and Structures) and the NIH under grant NIH R01EB029766.

Contents

List of Tables

Chapter 1
The Continuum Theory of Piezoelectricity and Piezomagnetism

1.1 Introduction

The focus of this book is on piezoelectricity and piezomagnetism in spatially random media. The piezoelectricity phenomenon is caused by the linear electromechanical interaction between the mechanical and electrical states in crystalline materials which lack inversion symmetry. In a piezoelectric material, an electrical charge is generated by the application of a mechanical force, while a mechanical deformation is caused by the application of an electrical field. Thus, from a thermodynamics standpoint, it is a reversible process. The situation is entirely analogous in a piezomagnetic material where, instead, the magnetic field plays the role of the electric field. Both piezoelectricity and piezomagnetism are caused by an absence of certain symmetries in a crystal structure.

While the theory of deterministic homogeneous piezoelectricity and piezomagnetism is classical, this book develops tools for random field theories. First, we generalise the continuum physics equations of inhomogeneous media and then develop a random field description of piezoelectric and piezomagnetic properties. Since in both cases, the key role is played by the same type of rank 3 tensor, we conduct the mathematical developments in terms of piezoelectricity.

The general approach in a continuum physics theory is that all interesting physical quantities are defined over a volume by the integration of density distributional functions, accounting for the effect of the same quantity defined for particles, which leads to the definition of each quantity with respect to spatial and time variables. These continuous physical quantities can then be employed to construct physical laws in terms of some integrals. Such a formulation enables us to fully characterise the macroscale behaviour of materials in the realm of thermodynamics. That is, one can develop a rational continuum thermodynamics in which the first and second laws of thermodynamics for a continuum matter are mathematically represented.

As mentioned earlier, at first, these laws have integral forms since their satisfaction needs to be addressed in the whole domain of the problem, resulting in some volume

A. Malyarenko et al., *Random Fields of Piezoelectricity and Piezomagnetism*,
SpringerBriefs in Mathematical Methods,
https://doi.org/10.1007/978-3-030-60064-8_1

and surface integrals. Next, by a localisation procedure, the aforementioned integral forms are reduced to a set of evolutionary partial differential equations (PDEs) along with some initial and boundary conditions. As the governing relations are not sufficient to fully characterise the physical quantities, more relations between the physical quantities are required. As we shall see below, this can be done through constitutive equations in which some of the dependent field variables are written in terms of the reset, so-called independent variables. Of course, we have various options in defining each variable to be dependent or independent, leading to different variants of governing equations. In addition, depending on whether the constitutive equations are linear or nonlinear—corresponding, respectively, to linear or nonlinear materials—we may arrive at linear or nonlinear PDEs. This is distinguished from another form of nonlinearity, called geometric nonlinearity, which is due to large deformations—a topic not treated in this book. Nevertheless, in practice, there are many situations in which the material behaviour is linear and the deformation gradient is small enough to justify the implementation of the linearised theory.

Clearly, the theory of classical continuum thermodynamics discussed above lacks coupling effects between thermal and electrical or electrical and mechanical states. In the early stages of development of rational continuum thermodynamics these effects seemed to be inconsequential. However, this is not the case nowadays as several experimental results refute this point of view. These materials are called multi-functional materials or smart materials and have found applications in devices such as ultrasonic transducers and micro-actuators, thermal-imaging devices, health monitoring devices, biomedical devices, and in biomimetics (bionics) and energy harvesting [19, 22]. As a result, it is necessary to develop a consistent rational mathematical model to analyse the behaviour of such materials. One way to accomplish this is to develop a coupled continuum framework in which the laws of classical continuum mechanics are augmented by a contribution due to the presence of electromagnetic fields.

This book includes material that has been developed after the publication of [20]. Therefore it extends it.

In the above book, the first- and second-named authors elaborated the general theory of spectral expansions of random fields taking values in a linear space of tensors of a fixed rank and, eventually, with a fixed index symmetry. As an example, they included random fields taking values in some subspaces of the 18-dimensional space of piezoelectricity tensors.

After the book was published, on the one hand, the second- and third-named authors developed a novel approach to the continuum theory of piezoelectricity and piezomagnetism which is applicable to random media. On the other hand, the first-named author found a simplified approach to spectral expansions of random fields that describe the above media. With this approach, a series of completely new examples of random fields that take values in the 3-dimensional space of the dihedral piezoelectricity class $[D_2]$, and which have various symmetries, from dihedral up to cubic, have been elaborated.

1.2 Continuum Electromagnetic Theory

This section summarises the foundations of continuum electromagnetic theory. We follow [8] both notionally and conceptually, referring the reader to the latter for more details.

To begin with, consider a deformable body whose volume, denoted by V, is a simply connected, open and bounded subset of three-dimensional Euclidean space, with boundary ∂V [12]. Furthermore, let $\overline{V} = V \cup \partial V$. Two different coordinate systems, namely the material and spatial coordinate systems, are used in this section. To distinguish between them, similarly to [8], bold and lowercase indices respectively stand for the material and spatial coordinates. Also, in the remainder of the chapter, standard index notation for Cartesian coordinates is occasionally used.

First, Maxwell's equations in a continuous matter, in a fixed frame, are [8]:

$$\begin{aligned} &\nabla \cdot \vec{B} = 0\,, \\ &\nabla \cdot \vec{D} = q_e\,, \\ &\nabla \times \vec{E} + \frac{1}{c}\frac{\partial \vec{B}}{\partial t} = \vec{0}\,, \\ &\nabla \times \vec{H} - \frac{1}{c}\frac{\partial \vec{D}}{\partial t} = \frac{1}{c}\vec{J}\,, \end{aligned} \tag{1.1}$$

where

$$\begin{aligned} &\vec{H} = \vec{B} - \vec{M}\,, \\ &\vec{D} = \vec{E} + \vec{P}\,, \end{aligned} \tag{1.2}$$

in which $\vec{E}$, $\vec{P}$, $\vec{D}$, $\vec{B}$, $\vec{M}$, $\vec{H}$, $\vec{J}$ and q_e are the electric field, volume electric polarisation, electric displacement field, magnetic induction, volume magnetic polarisation, magnetic field strength, electric current and charge density, respectively. It is worth mentioning that the above governing equations can be obtained through corresponding microscopic scales by means of statistical averaging (see [8] for details).

To generalise these equations for moving frames, which is needed for the balance laws of continuum mechanics in general, one must ensure that they are form-invariant under a Galilean transformation [8]:

$$\tilde{\vec{x}} = \vec{x} + \vec{V}t, \qquad \tilde{t} = t\,. \tag{1.3}$$

Here $\vec{V}$ is the constant velocity of the moving frame with respect to the fixed frame. When the velocity of the motion of the continuum is much less than c, the speed of light in vacuum, the necessary and sufficient conditions can be written as [8]:

$$
\begin{aligned}
\tilde{q}_e &= q_e \,, \\
\tilde{\vec{P}} &= \vec{P} \,, \\
\tilde{\vec{J}} &= \vec{J} + \vec{V} q_e \,, \\
\tilde{\vec{M}} &= \vec{M} - \frac{1}{c} \vec{V} \times \vec{P} \,, \\
\tilde{\vec{E}} &= \vec{E} - \frac{1}{c} \vec{V} \times \vec{B} \,, \\
\tilde{\vec{B}} &= \vec{B} + \frac{1}{c} \vec{V} \times \vec{E} \,, \\
\tilde{\vec{H}} &= \tilde{\vec{B}} - \tilde{\vec{M}} \,.
\end{aligned}
\tag{1.4}
$$

Consequently, if one assumes that at each time, say t, the velocity of the matter in the moving frame $R_C(\vec{x}, t)$ is $\vec{V} = -\vec{v}$, then we have:

$$
\begin{aligned}
\tilde{q}_e &= q_e \,, \\
\tilde{\vec{P}} &= \vec{P} \,, \\
\tilde{\vec{J}} &= \vec{J} - \vec{v} q_e \,, \\
\tilde{\vec{M}} &= \vec{M} + \frac{1}{c} \vec{v} \times \vec{P} \,, \\
\tilde{\vec{E}} &= \vec{E} + \frac{1}{c} \vec{v} \times \vec{B} \,, \\
\tilde{\vec{B}} &= \vec{B} - \frac{1}{c} \vec{v} \times \vec{E} \,, \\
\tilde{\vec{H}} &= \tilde{\vec{B}} - \tilde{\vec{M}} \,.
\end{aligned}
\tag{1.5}
$$

As a result, Maxwell's equations in the moving frame read [8]:

$$
\begin{aligned}
\nabla \cdot \vec{B} &= 0 \,, \\
\nabla \cdot \vec{D} &= q_e \,, \\
\nabla \times \tilde{\vec{E}} + \frac{1}{c} \vec{B}^* &= \vec{0} \,, \\
\nabla \times \tilde{\vec{H}} - \frac{1}{c} \vec{D}^* &= \frac{1}{c} \tilde{\vec{J}} \,,
\end{aligned}
\tag{1.6}
$$

where

$$
\begin{aligned}
\vec{B}^* &= \dot{\vec{B}} + \vec{B}(\nabla \cdot \vec{v}) - (\vec{B} \cdot \nabla)\vec{v} \,, \\
\vec{D}^* &= \dot{\vec{D}} + \vec{D}(\nabla \cdot \vec{v}) - (\vec{D} \cdot \nabla)\vec{v} \,,
\end{aligned}
\tag{1.7}
$$

are convective derivatives, which are frame indifferent, see [8, Definition (1.12.13)].

To consider the electromagnetic effects in a thermodynamic continuum, the balance laws—i.e., the balance of mass, the balance of linear momentum, the balance of angular momentum, the first law of thermodynamics, and the second law of thermodynamics—need to include a contribution due to electromagnetic fields. The approach to account for such contributions is quite similar to the case where a continuum theory is built by passing from particle mechanics. In other words, the macroscopic contribution due to electromagnetic fields—i.e., electromagnetic force, electromagnetic couple, and electromagnetic power—are derived from microscopic contributions by some statistical averaging (see [8] for details). In this way, after quite long manipulations, the generalised local form of balance laws can be written as [8]:

$$
\begin{aligned}
&\dot{\rho} + \rho v_{i,i} = 0\,, \\
&\rho \dot{v}_i = \sigma_{ji,j} + f_i + f_i^{em}\,, \\
&\sigma_{[ij]} = \widetilde{E}_{[i} P_{j]} + B_{[i} \widetilde{M}_{j]}\,, \\
&\rho \dot{e} = \sigma_{ij} v_{j,i} - q_{i,i} + \rho h + W_{in}^{em}\,, \\
-\rho(\dot{\psi} + \eta\dot{\theta}) + \sigma_{ij} v_{j,i} &+ \theta^{-1} q_i \theta_{,i} - P_i \widetilde{E}_{i}. - \widetilde{M}_i \dot{B}_i + \widetilde{J}_i \widetilde{E}_i \geq 0\,,
\end{aligned}
\tag{1.8}
$$

in which

$$
\begin{aligned}
\vec{f}^{em} &= q_e \widetilde{\vec{E}} + \frac{1}{c}(\widetilde{\vec{J}} + \vec{P}^*) \times \vec{B} + (\vec{P} \cdot \nabla)\widetilde{\vec{E}} + (\nabla \vec{B}) \cdot \widetilde{\vec{M}}\,, \\
W_{in}^{em} &= \vec{f}^{em}.\vec{v} + \rho \widetilde{\vec{E}} \cdot \dot{\vec{\pi}} - \widetilde{\vec{M}} \cdot \dot{\vec{B}} + \vec{J} \cdot \widetilde{\vec{E}}\,,
\end{aligned}
\tag{1.9}
$$

$$
\begin{aligned}
\vec{\pi} &= \frac{\vec{P}}{\rho}, \\
\vec{P}^* &= \dot{\vec{P}} + \vec{P}(\nabla \cdot \vec{v}) - (\vec{P} \cdot \nabla)\vec{v}\,,
\end{aligned}
$$

with electromagnetic jump or boundary conditions [8]:

$$
\begin{aligned}
\vec{n} \cdot \left[\vec{D}\right] &= 0\,, \\
\vec{n} \cdot \left[\vec{B}\right] &= 0\,, \\
\vec{n} \times \left[\widetilde{\vec{E}}\right] &= \vec{0}\,, \\
\vec{n} \times \left[\widetilde{\vec{H}}\right] &= \vec{0}\,,
\end{aligned}
\tag{1.10}
$$

with $\left[\vec{A}\right] \equiv \vec{A}^+ - \vec{A}^-$, where $\vec{A}^+$ and $\vec{A}^-$, respectively, denote the values of $\vec{A}$ when approaching from the positive and negative sides of a surface characterized by the normal vector $\vec{n}$ [8]. Zero surface current, zero surface free charge, and zero surface polarisation are assumed in (1.10) [8]. Also, σ_{ij}, $v_{i,j}$, q_i, v_i and $t_{i(n)}^{em}$ are, respectively, the components of Cauchy stress, gradient of velocity field, surface heat flux, velocity field, and electromagnetic surface traction. In addition, e, h, ψ, η and θ are the internal energy, rate of heat source per unit volume, Helmholtz free energy density, entropy, and temperature, respectively. Furthermore, $\sigma_{[ij]}$ denotes the skew-symmetric part of the Cauchy stress tensor, i.e., $\sigma_{[ij]} = \frac{1}{2}(\sigma_{ij} - \sigma_{ji})$. Clearly, the total stress in the presence of an electromagnetic field is no longer symmetric.

Equations (1.1) and (1.8), together with boundary conditions, give the governing equations of a deformable body in the presence of electromagnetic fields. However, similar to the case of classical continuum mechanics, one needs to further define constitutive equations to successfully resolve the intrinsic indeterminate nature of the problem. In this regard, the constitutive equations need to satisfy several axioms that we, herein, just enumerate as follows [8]: (1) causality, (2) determinism, (3) equipresence, (4) objectivity, (5) time reversal, (6) material invariance, (7) neighbourhood, (8) memory, and (9) admissibility. The systematic approach to obtaining constitutive equations is to define a sufficient number of functionals, say $\Im^{\alpha}$, that meet the aforementioned axioms. To this end, following [8], the dependent physical quantities are taken as follows:

$$\mathcal{L} = \left\{\vec{T}^{E}, \vec{Q}, \Psi, \eta, \Pi, \vec{M}, \vec{J}\right\}, \tag{1.11}$$

while the independent ones are:

$$\mathcal{Y} = \left\{\vec{\widetilde{E}}, \theta, \vec{B}\right\} \tag{1.12}$$

along with spatial coordinates and a time parameter. Equations (1.11) and (1.12) are represented in material coordinates, which are related to spatial coordinates via the following relations [8]:

$$\begin{aligned} &\rho = J\rho_0, \quad T^{E}_{IJ} = JX_{I,i}X_{J,j}\,\sigma_{ij}\,, \\ &Q_I = JX_{I,i}q_i, \Pi_I = JX_{I,i}P_i, \quad M_I = JX_{I,i}\widetilde{M}_i, \quad \widetilde{J}_I = JX_{I,i}J_i, \quad J = \det x_{i,I}\,. \end{aligned} \tag{1.13}$$

$\vec{T}^{E}$, in (1.11), is the symmetric tensor in the reference configuration corresponding to $\vec{\sigma}$. Next, the first and second laws of thermodynamics in material coordinates can be formulated in the following form [8]:

$$\begin{aligned} &\rho_0(\dot{\psi} + \dot{\theta}\eta + \dot{\eta}\theta) = \frac{1}{2}T^{E}_{IJ}\dot{C}_{IJ} + Q_{I,I} - \rho_0 h + \Pi_I\,\dot{\widetilde{E}}_I - M_I\dot{B}_I + \widetilde{J}_I\widetilde{E}_I\,, \\ &-\rho_0(\dot{\psi} + \dot{\theta}\eta) + \frac{1}{2}T^{E}_{IJ}\dot{C}_{IJ} + \frac{1}{\theta}Q_I\theta_I - \Pi_I\,\dot{\widetilde{E}}_I - M_I\dot{B}_I + \widetilde{J}_I\widetilde{E}_I \geq 0\,. \end{aligned} \tag{1.14}$$

It is easy to see that 20 more functionals are required to have the same number of unknowns and equations. In doing so, we define the Helmholtz free energy for a deformable solid with heat, electrical, and mechanical coupling as follows [8]:

$$\rho_0\psi = \tilde{\psi}(\vec{C}, \tilde{\vec{E}}, \tilde{\vec{B}}, \nabla_R\theta; \theta, \vec{X})\,. \tag{1.15}$$

By taking the derivative, we have:

$$\rho_0\dot{\psi} = \frac{\partial\tilde{\psi}}{\partial C_{IJ}}\dot{C}_{IJ} + \frac{\partial\tilde{\psi}}{\partial\tilde{E}_I}\dot{\tilde{E}}_I + \frac{\partial\tilde{\psi}}{\partial B_I}\dot{\tilde{B}}_I + \frac{\partial\tilde{\psi}}{\partial\theta_{,I}}\dot{\theta}_{,I} + \frac{\partial\tilde{\psi}}{\partial\theta}\dot{\theta}\,. \tag{1.16}$$

Substituting the result into (1.14) and considering that $\dot{\theta}$, $\dot{C}_{IJ}$, $\dot{\theta}_{,I}$, $\dot{\tilde{E}}_i$ and $\dot{B}_I$ are independent variables, the inequality holds when the following relations are true [8]:

$$\eta = -\frac{1}{\rho_0}\frac{\partial\tilde{\psi}}{\partial\theta},\; T^E_{IJ} = 2\frac{\partial\tilde{\psi}}{\partial C_{IJ}},\; \frac{\partial\tilde{\psi}}{\partial\theta_{,I}} = 0,\; \Pi_I = -\frac{\partial\tilde{\psi}}{\partial\tilde{E}_I},\; M_I = -\frac{\partial\tilde{\psi}}{\partial\tilde{B}_I}, \tag{1.17}$$

with [8]:

$$\begin{aligned} Q_I &= \tilde{Q}_I(\vec{C},\ \tilde{\vec{E}},\ \vec{B},\ \nabla_R\theta;\ \theta,\ \vec{X}),\\ \tilde{J}_I &= \tilde{J}_I(\vec{C},\ \tilde{\vec{E}},\ \vec{B},\ \nabla_R\theta;\ \theta,\ \vec{X})\,. \end{aligned} \tag{1.18}$$

Observe that (1.18) satisfies [8]:

$$\frac{1}{\theta}Q_{,I}\theta_{,I} + J_I\tilde{E}_I \geq 0\,. \tag{1.19}$$

Up to this point, all the equations have been written in general form, i.e., no assumption has been made within the continuum framework. However, in many applications, a linearised theory is sufficient to find accurate results. To this end, let us assume that the natural state is free of fields, that is, it has the following form [21]:

$$x^0 = X,\; \Pi^0_I = 0,\; E^0_{IJ} = 0,\; E^0_I = 0,\; B^0_I = 0, \ldots\,. \tag{1.20}$$

In addition, suppose that all the fields are small enough in the sense that [8, 21]

$$\begin{aligned} &\hat{\vec{u}} = \delta\vec{u},\ \hat{\vec{E}} = \delta\vec{E},\ \hat{\vec{B}} = \delta\vec{B},\ \hat{\Pi} = \delta\Pi\,,\\ &\hat{\theta} - \hat{\theta}_0 = \delta\theta,\ \left|\delta\hat{\theta}\right| << \hat{\theta}_0,\ \hat{\theta}_0 > 0,\ \nabla\hat{\theta} = \delta\nabla\theta\ldots\,, \end{aligned} \tag{1.21}$$

where δ is a small number. Linearising nonlinear fields by expansion about the natural state and neglecting nonlinear terms, we find:

$$
\begin{aligned}
&\rho\ddot{u}_i = \sigma_{ij,j} + f_i\,,\\
&\sigma_{ij}n_j = t_{(n)i}\,,\\
&\sigma_{[ij]} = 0\,,\\
&\rho\dot{e} = \sigma_{ij}v_{i,j} - q_{i,i} + \rho h + W_{in}^{em}\,,\\
&W_{in}^{em} = J_i E_i + E_i\frac{\partial P_i}{\partial t} - M_i\frac{\partial B_i}{\partial t}\,,\\
-&\rho(\dot{\psi} + \eta\dot{\theta}) + \sigma_{ij}v_{i,j} + \theta^{-1}q_i\theta_{,i} - P_i\dot{E}_i - M_i\dot{B}_i + J_i E_i \geq 0\,,
\end{aligned}
\tag{1.22}
$$

while Maxwell's equations remain unchanged. Furthermore, due to the fact that all fields are small, a set of linear constitutive equations is sufficient. Hence, as a consequence of (1.16), $\tilde{\psi}$ has a quadratic form as follows [8]:

$$
\begin{aligned}
\tilde{\psi} = \tilde{\psi}_0 + \tilde{\psi}_{IJ}\varepsilon_{IJ} + \frac{1}{2}C_{IJKL}\varepsilon_{IJ}\varepsilon_{KL} - D_{IJK}^E E_I\varepsilon_{JK} - D_{IJK}^B B_I\varepsilon_{JK} - \chi_I^E E_I\\
-\frac{1}{2}\chi_{IJ}^E E_I E_J - \chi_I^B B_I - \frac{1}{2}\chi_{IJ}^B B_I B_J - \Lambda_{IJ}E_I B_J\,,
\end{aligned}
\tag{1.23}
$$

in which $\tilde{\psi}_0, \tilde{\psi}_{IJ},\ C_{IJKL},\ D_{IJK}^E,\ D_{IJK}^B,\ \chi_I^E,\ \chi_{IJ}^E,\ \chi_I^B,\ \chi_{IJ}^B$ and Λ_{IJ} are functions of the $\vec{X}$ and θ. Here ϵ_{IJ} denotes an infinitesimal strain tensor. By making use of the assumption given in (1.19), $\tilde{\psi}_0, \tilde{\psi}_{IJ},\ \chi_I^E$ and χ_I^B in (1.23) can be expressed in terms of temperature in the following form [8]:

$$
\begin{aligned}
&\tilde{\psi}_0 = -\rho_0\eta_0\theta - (\frac{\rho_0\gamma}{2\theta_0})\theta^2, \quad \tilde{\psi}_{IJ} = -B_{IJ}\theta\,,\\
&\chi_I^E = \tilde{\omega}_I\theta, \quad \chi_I^B = \Gamma_I\theta\,.
\end{aligned}
\tag{1.24}
$$

Hence, based on (1.17), one can derive [8]:

$$
\begin{aligned}
\eta &= \eta_0 + \frac{\gamma}{\theta_0}\theta + \frac{1}{\rho_0}(B_{IJ}\varepsilon_{IJ} + \widetilde{\omega}_I E_I + \Gamma_I B_I)\,,\\
T_{IJ}^E &= -B_{IJ}\theta + C_{IJKL}\varepsilon_{KL} - D_{KIJ}^E E_K - D_{KIJ}^B B_K\,,\\
\Pi_I &= \widetilde{\omega}_I\theta + \chi_{IJ}^E E_J + D_{IJK}^E\varepsilon_{JK} + \Lambda_{IJ}B_J\,,\\
M_I &= \Gamma_I\theta + \chi_{IJ}^B B_J + D_{IJK}^B\varepsilon_{JK} + \Lambda_{JI}E_J\,.
\end{aligned}
\tag{1.25}
$$

Analogously, by taking into account the Clausius–Duhem inequality, (1.18) can be written in the form [8]:

$$
\begin{aligned}
Q_I &= \kappa_{IJ}\theta_{,J} + \kappa_{IJ}^E\widetilde{E}_J\,,\\
\widetilde{J}_I &= \sigma_{IJ}\widetilde{E}_J + \sigma_{IJ}^\theta\theta_{,J}\,,
\end{aligned}
\tag{1.26}
$$

where κ_{IJ} and σ_{IJ} are symmetric and semi-positive definite. Also, let us assume that the material and spatial frames coincide. Therefore, we have:

$$
x_i \approx X_J\delta_{iJ}\,,
\tag{1.27}
$$

where $\delta_{iJ} = \delta_{ij} = \delta_{Ij} = \delta_{IJ}$. Hence, in linear theory, there is no difference between the material and the spatial frame and each description can be interchangeably used.

Now, as was mentioned earlier, one can derive various constitutive equations based on independent and dependent variables chosen in the problem. In the linear case, nevertheless, by employing an appropriate Legendre transformation, it is straightforward to obtain all possible forms from one form. In the following, we shall consider these constitutive equations [27]:

$$\begin{aligned}
\varepsilon_{ij} &= S_{ijkl}\sigma_{kl} + d^{E}_{kij}E_k + d^{H}_{kij}H_k + \alpha_{ij}\theta \,, \\
D_i &= d^{E}_{ikl}\sigma_{kl} + \chi^{E}_{ij}E_j + \chi^{EH}_{ij}H_j + \beta^{E}_{i}\theta \,, \\
B_i &= d^{H}_{ikl}\sigma_{kl} + \chi^{EH}_{ji}E_j + \chi^{H}_{ij}H_j + \beta^{H}_{i}\theta \,, \\
\eta &= \alpha_{ij}\sigma_{ij} + \beta^{E}_{i}E_i + \beta^{H}_{i}H_i + \alpha^{s}\theta \,,
\end{aligned} \tag{1.28}$$

and

$$\begin{aligned}
\sigma_{ij} &= C_{ijkl}\varepsilon_{kl} - e^{E}_{kij}E_k - e^{H}_{kij}H_k + \alpha_{ij}\theta \,, \\
D_i &= e^{E}_{ikl}\varepsilon_{kl} + \kappa^{E}_{ij}E_j + \kappa^{EH}_{ij}H_j + \beta^{E}_{i}\theta \,, \\
B_i &= e^{H}_{ikl}\varepsilon_{kl} + \kappa^{EH}_{ji}E_j + \kappa^{H}_{ij}H_j + \beta^{H}_{i}\theta \,, \\
\eta &= \alpha_{ij}\varepsilon_{ij} + \beta^{E}_{i}E_i + \beta^{H}_{i}H_i + \alpha^{s}\theta \,,
\end{aligned} \tag{1.29}$$

in which, and hereafter, we use lowercase letters for indices. Clearly, each constitutive coefficient given in (1.28) and (1.29) has a physical meaning; S_{ijkl}, C_{ijkl}, d^{E}_{kij}, d^{H}_{kij}, e^{E}_{kij}, e^{H}_{kij}, χ^{E}_{ij}, χ^{H}_{ij}, κ^{E}_{ij}, κ^{H}_{ij}, χ^{EH}_{ij} and κ^{EH}_{ij} denote components of the compliance tensor, stiffness tensor, direct piezoelectric tensor, direct piezomagnetic tensor, reverse piezoelectric tensor, reverse piezomagnetic tensor, permittivity under constant stress, permeability under constant stress, permittivity under constant strain, permeability under constant strain, magnetoelectric tensor under constant stress and magnetoelectric tensor under constant strain, respectively [3].

1.3 The Theory of Linear Piezoelectric Materials

So far, we have summarised the continuum electromagnetic theory for both linear and nonlinear regimes. In the following, we focus on the theory of linear elastic dielectrics, i.e., materials that are not electrically conducting, while their magnetic property is negligible [8]. Also, for the sake of simplicity, we only consider the case of the isothermal condition. In this regard, analogous to the case of classical continuum mechanics, there shall be two approaches to analyse such materials: (i) the Displacement approach; and (ii) the Stress approach [25]. The reason behind each method lies mainly in the type of prescribed boundary conditions. As will be seen, to theoretically obtain a particular solution for a dielectric body, both the prescribed boundary conditions and the initial conditions are required. It turns out that the displacement approach is a more suitable method when the prescribed boundary conditions are of a mixed-type, that is, when on some part of the boundary the displacement is prescribed, while on the remainder the traction is prescribed. Of course, having a displacement field prescribed over the entire boundary is a special case of the category mentioned above. On the other hand, when the mechanical boundary conditions are traction-type, the stress approach is more desirable as both governing equations and boundary conditions are written in terms of the stress field. As an example, from a computational physics perspective, by making use of the stress governing equations, one can develop a stress-type finite element method in which the exact satisfaction of boundary conditions is assured. Hence, in the remainder of the present chapter, we explain both methods, so that the reader can appreciate the regular types of PDEs we typically encounter in the analysis of linear piezoelectric materials. In addition, at the end of the chapter, we consider a special category of variational principles—namely, convolution variational forms—as an alternative means of constructing numerical methods, e.g., finite element methods, for the analysis of such materials.

It should be mentioned that, from now on, body force per unit mass is used. Let us first list the governing equations relevant to the linear piezoelectric case under the isothermal condition. To do so, we assume the quasi-electrostatic condition, which is applicable in a number of situations, e.g., ultrasonics:

$$\begin{aligned} &\nabla \cdot \vec{\sigma} + \rho \vec{f} = \rho \ddot{\vec{u}} \quad \text{on } V \times (0, \infty)\,, \\ &\nabla \cdot \vec{D} = q_e \quad \text{on } V \times (0, \infty)\,, \\ &\nabla \times \vec{E} = 0 \quad \text{on } V \times (0, \infty)\,. \end{aligned} \tag{1.30}$$

It follows that the electric field can be expressed in terms of the gradient of some scalar field, say an electric potential function:

$$E_i = -\varphi_{,i}(\vec{x}, t) \text{ on } V \times (0, \infty)\,. \tag{1.31}$$

Now, as a result of linearity, we have the following kinematic condition:

$$\varepsilon_{ij} = \frac{1}{2}(u_{i,j} + u_{j,i}) \quad \text{on } V \times (0, \infty)\,. \tag{1.32}$$

As there exist two sets of governing equations, we naturally have two types of boundary conditions: (i) mechanical boundary conditions; and (ii) electrical boundary conditions. Hence, partitioning the boundary ∂V into the following complementary subsets:

$$\begin{aligned} &\partial V = \partial V_{\vec{u}} \cup \partial V_{\vec{\sigma}}\,, \quad \partial V_{\vec{u}} \cap \partial V_{\vec{\sigma}} = \varnothing\,, \\ &\partial V = \partial V_{\varphi} \cup \partial V_{\vec{D}}\,, \quad \partial V_{\varphi} \cap \partial V_{\vec{D}} = \varnothing\,, \end{aligned} \tag{1.33}$$

we define the boundary conditions as follows:

$$
\begin{aligned}
&\text{Mechanical Boundary Conditions} \quad \begin{cases} u_i = \hat{u}_i(\vec{x}, t) & \text{on } \partial V_{\vec{u}} \times [0, \infty)\,, \\ \sigma_{ij} n_j = \hat{t}_i(\vec{x}, t) & \text{on } \partial V_{\vec{\sigma}} \times [0, \infty)\,, \end{cases} \\
&\text{Electrical Boundary Conditions} \quad \begin{cases} \varphi = \hat{\varphi}(\vec{x}, t) & \text{on } \partial V_{\varphi} \times [0, \infty)\,, \\ D_j n_j = \hat{d}(\vec{x}, t) & \text{on } \partial V_{\vec{D}} \times [0, \infty)\,, \end{cases}
\end{aligned} \tag{1.34}
$$

in which $\hat{u}_i(\vec{x}, t)$, $\hat{t}_i(\vec{x}, t)$, $\hat{\varphi}(\vec{x}, t)$ and $\hat{d}(\vec{x}, t)$ are, respectively, the prescribed displacement vector, traction vector, electric potential and electric displacement over the boundary. Moreover, the associated initial conditions are:

$$
\begin{aligned}
u_i(\vec{x}, 0) &= u_i^0(\vec{x}), \;\; \vec{x} \in \overline{V}, \\
\dot{u}_i(\vec{x}, 0) &= \dot{u}_i^0(\vec{x}), \;\; \vec{x} \in \overline{V}.
\end{aligned} \tag{1.35}
$$

As mentioned earlier, in the linear regime, there are variants of constitutive equations that can be obtained from one another by employing the Legendre transformation. Since it is desired to obtain displacement and stress approaches, we utilise the following forms [3]:

$$
\begin{aligned}
\sigma_{ij} &= C_{ijkl}\varepsilon_{kl} - e^E_{kij} E_k \;\; \text{on } V \times (0, \infty)\,, \\
D_i &= e^E_{ikl}\varepsilon_{kl} + \kappa^E_{ij} E_j \;\; \text{on } V \times (0, \infty)\,,
\end{aligned} \tag{1.36}
$$

and alternatively

$$
\begin{aligned}
\varepsilon_{ij} &= S_{ijkl}\sigma_{kl} + d^E_{kij} E_k \;\; \text{on } V \times (0, \infty)\,, \\
D_i &= d^E_{ikl}\sigma_{kl} + \chi^E_{ij} E_j \;\; \text{on } V \times (0, \infty)\,.
\end{aligned} \tag{1.37}
$$

The following symmetry conditions hold [3]:

$$
\begin{aligned}
C_{ijkl} &= C_{klij} = C_{jikl} = C_{ijlk} \;\; \text{on } V\,, \\
e^E_{kij} &= e^E_{kji} \;\; \text{on } V\,, \\
\kappa^E_{ij} &= \kappa^E_{ji} \;\; \text{on } V\,, \\
S_{ijkl} &= S_{klij} = S_{jikl} = S_{ijlk} \;\; \text{on } V\,, \\
d^E_{kij} &= d^E_{kji} \;\; \text{on } V\,, \\
\chi^E_{ij} &= \chi^E_{ji} \;\; \text{on } V\,,
\end{aligned} \tag{1.38}
$$

with the following relations:

$$
\begin{aligned}
S_{ijkl}C_{ijpq} &= \delta_{kp}\delta_{lq} \;\; \text{on } V\,, \\
d^E_{kij} &= S_{pqij}e^E_{kpq} \;\; \text{on } V\,, \\
\chi^E_{ij} &= S_{pqrs}e^E_{ipq}e^E_{jrs} + \kappa^E_{ij} \;\; \text{on } V\,.
\end{aligned} \tag{1.39}
$$

Also, as will be employed in the final section of this chapter, let us specifically define what we mean by an admissible piezoelectric process.

Definition 1.1 An ordered array $\Omega = [\vec{u}, \vec{\varepsilon}, \vec{\sigma}, \vec{E}, \vec{D}, \varphi]$ is called a *piezoelectric process* provided that

$$\begin{aligned} &u_i \in C^{1,2}, \quad \varepsilon_{ij} \in C^{0,0}, \quad \sigma_{ij} \in C^{1,0}, \quad D_i \in C^{1,0}, \quad \varphi \in C^{1,0}, \\ &E_i \in C^{0,0}, \quad \varepsilon_{ij} = \varepsilon_{ji}, \quad \sigma_{ij} = \sigma_{ji}, \end{aligned} \tag{1.40}$$

where by $C^{M,N}$ we mean the set of all functions whose spatial and time derivatives up to the order, respectively, M and N exist and are continuous. In this sense, an admissible process that satisfies (1.30), (1.31), (1.32), (1.34), (1.35) and (1.36) is a solution of a mixed initial boundary value problem governing the motion of a piezoelectric material under the aforementioned assumptions.

To proceed further, let us assume the following continuity conditions [3]:

i. $\rho > 0$ is continuously differentiable on $\overline{V}$;
ii. $\vec{C}$, $\vec{e}^E$, $\vec{\kappa}^E$ and $\vec{S}$, $\vec{d}^E$, $\vec{\chi}^E$ are continuously differentiable on $\overline{V}$ and admit (1.38), respectively;
iii. $\vec{u}^0(\vec{x})$ is continuously differentiable on $\overline{V}$;
iv. $\vec{v}^0(\vec{x})$ is continuously differentiable on $\overline{V}$;
v. $\vec{f}$ and q_e are continuously differentiable on $\overline{V}$;
vi. $\hat{\vec{u}}$ and $\hat{\varphi}$ are continuous on $\overline{\partial V_{\vec{u}}} \times [0, \infty)$ and $\overline{\partial V_{\vec{D}}} \times [0, \infty)$, respectively;
vii. $\hat{t}$ and $\hat{d}$ are piecewise continuous on $\overline{\partial V_{\vec{\sigma}}} \times [0, \infty)$ and $\overline{\partial V_{\vec{D}}} \times [0, \infty)$, respectively.

1.3.1 The Displacement Approach

By making use of the above-mentioned equations, one can write the governing equations in terms of displacements and electric potential. Substituting (1.31) and (1.32) into (1.36), we find:

$$\begin{aligned} \sigma_{ij} &= C_{ijkl} u_{k,l} + e^E_{kij} \varphi_{,k}, \\ D_i &= e^E_{ikl} u_{k,l} - \kappa^E_{ij} \varphi_{,j}, \end{aligned} \tag{1.41}$$

where the symmetry condition (1.38) has been employed.

The following theorem represents a set of governing equations in terms of displacement and electric potential:

Theorem 1.1 *Let $\Omega = [\vec{u}, \varphi]$ denote an ordered array with $\vec{u} \in C^{2,2}$ and $\varphi \in C^{2,0}$. Then Ω is a solution of the mixed initial boundary value problem if and only if:*

$$
\begin{aligned}
&(C_{ijkl}u_{k,l} + e^E_{kij}\varphi_{,k})_{,j} + \rho f_i = \rho \ddot{u}_i \; on \; V \times (0, \infty)\,, \\
&(e^E_{ikl}u_{k,l} - \kappa^E_{ij}\varphi_{,j})_{,i} = q_e \; on \; V \times (0, \infty)\,, \\
&u_i = \hat{u}_i(\vec{x}, t) \; on \; \partial V_{\vec{u}} \times [0, \infty)\,, \\
&(C_{ijkl}u_{k,l} + e^E_{kij}\varphi_{,k})n_j = \hat{t}_i(\vec{x}, t) \; on \; \partial V_{\vec{\sigma}} \times [0, \infty)\,, \\
&\varphi = \hat{\varphi}(\vec{x}, t) \; on \; \partial V_{\varphi} \times [0, \infty)\,, \\
&(e^E_{ikl}u_{k,l} - \kappa^E_{ij}\varphi_{,j})n_i = \hat{d}(\vec{x}, t) \; on \; \partial V_{\vec{D}} \times [0, \infty)\,, \\
&u_i(\vec{x}, 0) = u^0_i(\vec{x}), \;\; \vec{x} \in \bar{V}, \\
&\dot{u}_i(\vec{x}, 0) = \dot{u}^0_i(\vec{x}), \;\; \vec{x} \in \bar{V}\,.
\end{aligned}
\tag{1.42}
$$

The previous theorem gives us the governing equations by which one can find a solution to the mixed initial boundary value problem, but no information as to whether or not the solution is unique. In the following statement, we give a sufficient condition for the uniqueness of the solution [8]:

Theorem 1.2 *Let V be a regular region. Let $\vec{C}$ and $\vec{\chi}^E$ be, respectively, a positive definite fourth-order and second-order tensor. Then there is at most one solution for the mixed initial-boundary value problem.*

1.3.2 The Stress Approach

In this section, we discuss the governing equations in terms of the stress field. This approach was originally employed as an alternative framework in [14, 15]. Proceeding in that vein, [11, 12] generalised the method, so it could be applied to the case of a mixed boundary condition, leading to a set of governing equations that are of integro-differential type. In addition, some special forms of the variational formulation have been derived in [11, 12]. These forms, in contrast to the classical variational forms, guarantee the systematic satisfaction of initial conditions. Subsequently, various studies have been carried out to explore the potential of the stress language (approach), see the review [25]. To obtain the governing equations, let us take the Laplace transform of $(1.30)_1$:

$$
\rho \ell[u_i] = \rho \left(\frac{u^0_i}{s} + \frac{\dot{u}^0_i}{s^2} \right) + \frac{\ell[\sigma_{ij,j} + f_i]}{s^2}\,. \tag{1.43}
$$

After some manipulation we obtain:

$$
\begin{aligned}
&\rho u_i = \rho(u^0_i + t\,\dot{u}^0_i) + t * (\sigma_{ij,j} + f_i) \\
&\text{or} \\
&\rho \vec{u} = t * \nabla.\vec{\sigma} + t * \vec{f} + \rho(\vec{u}^0 + t\,\dot{\vec{u}}^0)\,,
\end{aligned}
\tag{1.44}
$$

in which $*$ denotes the convolution product of two functions in the sense of:

$$f * g(\vec{x}, t) = \int_0^t f(\vec{x}, t - \lambda) g(\vec{x}, \lambda) \mathrm{d}\lambda \quad (\vec{x}, t) \in \Re \times [0, \infty) , \tag{1.45}$$

where $\Re$, in general, stands for a subset of Euclidean space. Thus, we have the following theorems [12]:

Theorem 1.3 *Let $u_i \in C^{0,2}$ and $\sigma_{ij} \in C^{1,0}$ be a vector field and a second-order symmetric tensor field, respectively. Then $\vec{u}$ and $\vec{\sigma}$ meet $(1.30)_1$ and the associated initial conditions if and only if*

$$\rho \vec{u} = t * \nabla . \vec{\sigma} + t * \vec{f} + \rho (\vec{u}^0 + t \dot{\vec{u}}^0) . \tag{1.46}$$

It is worth mentioning that the remaining results presented in this section are a reduction of those originally derived in [3] for the case of electro-magneto-elastic materials. Alternative conditions for obtaining a solution are thus:

Theorem 1.4 *The admissible process $\Omega = [\vec{u}, \vec{\varepsilon}, \vec{\sigma}, \vec{E}, \vec{D}, \varphi]$ is a solution of the initial mixed BVP if and only if it satisfies $(1.30)_2$, (1.31), (1.32), (1.34), (1.36) and (1.46).*

Now, suppose that $\partial V_{\vec{u}} = \varnothing$. Let us alternatively have

$$\begin{aligned} \sigma_{ij}(\vec{x}, 0) &= \sigma_{ij}^0(\vec{x}), \; \varphi(\vec{x}, 0) = \varphi^0(\vec{x}), \; \vec{x} \in \bar{V} , \\ \dot{\sigma}_{ij}(\vec{x}, 0) &= \dot{\sigma}_{ij}^0(\vec{x}), \; \dot{\varphi}(\vec{x}, 0) = \dot{\varphi}^0(\vec{x}), \; \vec{x} \in \bar{V} , \end{aligned} \tag{1.47}$$

in which $\sigma_{ij}^0(\vec{x})$, $\dot{\sigma}_{ij}^0(\vec{x})$, $\varphi^0(\vec{x})$, and $\dot{\varphi}^0(\vec{x})$ are the corresponding prescribed initial conditions for the problem being consistent in the sense of

$$\begin{aligned} S_{ijkl}\sigma_{kl}^0(\vec{x}) - d_{kij}^E \varphi_{,k}^0(\vec{x}) &= \frac{1}{2}(u_{i,j}^0 + u_{j,i}^0), \quad \vec{x} \in \bar{V} , \\ S_{ijkl}\dot{\sigma}_{kl}^0(\vec{x}) - d_{kij}^E \dot{\varphi}_{,k}^0(\vec{x}) &= \frac{1}{2}(\dot{u}_{i,j}^0 + \dot{u}_{j,i}^0), \quad \vec{x} \in \bar{V} , \end{aligned} \tag{1.48}$$

and which, by using (1.34) and (1.47), can be uniquely obtained from (1.48).

Lemma 1.1 *Let $u_i(\vec{x}, t)$, $\varepsilon_{ij}(\vec{x}, t)$, $\sigma_{ij}(\vec{x}, t)$, $E_i(\vec{x}, t)$ and $\varphi(\vec{x}, t)$ satisfy (1.31), (1.32), $(1.36)_1$ and (1.46) with*

$$\vec{\sigma} \in C^{2,2} \text{ on } \bar{V} \times [0, \infty), \; \varphi \in C^{2,2} \text{ on } \bar{V} \times [0, \infty), \; \sigma_{ij} = \sigma_{ji} . \tag{1.49}$$

Then, we have:

$$S_{ijkl}\ddot{\sigma}_{kl} = \left(\frac{1}{\rho}\sigma_{ik,k}\right)_{,j} + \left(\frac{1}{\rho} f_i\right)_{,j} + d_{kij}^E \ddot{\varphi}_{,k} . \tag{1.50}$$

Lemma 1.2 *Suppose (1.49) holds. Define u_i by (1.46), ε_{ij} by (1.37)$_1$, and E_i by (1.31). Let $\vec{\sigma}$, $\vec{\sigma} = \vec{\sigma}^T$, and φ satisfy (1.50). Then the kinematic equation (1.32) holds true.*

Now, with the aid of Lemmas 1.1 and 1.2, one can obtain the following theorem:

Theorem 1.5 *Let $\vec{\sigma} \in C^{2,2}$, $\varphi \in C^{2,2}$ with $\sigma_{ij} = \sigma_{ji}$. Then an ordered array $\Omega = [\vec{\sigma}, \varphi]$ is a solution of the initial mixed BVP if and only if:*

$$
\begin{aligned}
&S_{ijkl}\ddot{\sigma}_{kl} = (\frac{1}{\rho}\sigma_{ik,k})_{,j} + (\frac{1}{\rho}f_i)_{,j} + d^E_{kij}\ddot{\varphi}_{,k} \text{ on } V \times (0,\infty)\,,\\
&(d^E_{ikl}\sigma_{kl} - \chi^E_{ij}\varphi_{,j})_{,i} = q_e \text{ on } V \times (0,\infty)\,,\\
&\sigma_{ij}n_j = \hat{t}_i(\vec{x},t) \text{ on } \partial V \times [0,\infty)\,,\\
&\varphi = \hat{\varphi}(\vec{x},t) \text{ on } \partial V_\varphi \times [0,\infty)\,,\\
&(d^E_{ikl}\sigma_{kl} - \chi^E_{ij}\varphi_{,j})n_i = \hat{d}(\vec{x},t) \text{ on } \partial V_{\vec{D}} \times [0,\infty)\,,\\
&\sigma_{ij}(\vec{x},0) = \sigma^0_{ij}(\vec{x}),\ \varphi(\vec{x},0) = \varphi^0(\vec{x}),\ \vec{x} \in \bar{V}\,,\\
&\dot{\sigma}_{ij}(\vec{x},0) = \dot{\sigma}^0_{ij}(\vec{x}),\ \dot{\varphi}(\vec{x},0) = \dot{\varphi}^0(\vec{x}),\ \vec{x} \in \bar{V}\,,\\
&u_i(\vec{x},0) = u^0_i(\vec{x}),\ \vec{x} \in \bar{V}\,,\\
&\dot{u}_i(\vec{x},0) = \dot{u}^0_i(\vec{x}),\ \vec{x} \in \bar{V}\,.
\end{aligned}
\tag{1.51}
$$

1.4 Variational Principles

In this section, we provide convolutional variational principles concerning piezoelectric materials. These principles are interesting from the viewpoint of establishing numerical methods, e.g., finite element methods, to solve practical problems that often involve complicated geometry and boundary and initial conditions. In this regard, the following results are special cases of those of [4] for the case of electro-magneto-elastic materials.

To begin with, we have the following definition [4, 12]

Definition 1.2 Let L be a linear space, K a subset of L, and $\Phi(S)$ a functional on K. We define

$$
\delta_{\tilde{S}}\,\Phi(S) = \frac{\mathrm{d}}{\mathrm{d}\lambda}\Phi(S + \lambda\tilde{S})\bigg|_{\lambda=0}
\tag{1.52}
$$

for all real numbers λ, where $S,\ \tilde{S} \in L$ and $S + \lambda\tilde{S} \in K$, and we say the variation of $\Phi(S)$ is zero and write '$\delta\,\Phi(S) = 0$ over L' if and only if $\delta_{\tilde{S}}\,\Phi(S)$ exists and is equal to zero for all $\tilde{S}$ such that $S,\ \tilde{S} \in L$ and $S + \lambda\tilde{S} \in K$.

The following theorem represents a variational form in which no restriction on the desired quantities is assumed:

Theorem 1.6 *Let Ω be the set of all admissible processes. Let $S = [\vec{u}, \vec{\varepsilon}, \vec{\sigma}, \vec{E}, \vec{D}, \varphi]$ be an element of Ω. Define the functional ϑ_t on Ω at each time, say $t \in [0, \infty)$, by*

$$
\begin{aligned}
\vartheta_t(S) =& \frac{1}{2}\int_V C_{ijkl}(\vec{x})\left[h * \varepsilon_{ij} * \varepsilon_{kl}\right](\vec{x}, t)\mathrm{d}\vec{x} - \int_V \left[h * \sigma_{ij} * \varepsilon_{ij}\right](\vec{x}, t)\mathrm{d}\vec{x} \\
& - \int_V e^E_{kij}(\vec{x})\left[h * \varepsilon_{ij} * E_k\right](\vec{x}, t)\mathrm{d}\vec{x} - \frac{1}{2}\int_V \kappa^E_{ij}(\vec{x})\left[h * E_i * E_j\right](\vec{x}, t)\mathrm{d}\vec{x} \\
& + \int_V [h * D_i * E_i](\vec{x}, t)\mathrm{d}\vec{x} - \int_V [h * (D_{i,i} - q_e) * \varphi](\vec{x}, t)\mathrm{d}\vec{x} \\
& + \frac{1}{2}\int_V \rho(\vec{x})\,[u_i * u_i]\,(\vec{x}, t)\mathrm{d}\vec{x}n - \int_V \left[(h * \sigma_{ij,j} + f^b_i) * u_i\right](\vec{x}, t)\mathrm{d}\vec{x} \\
& + \int_{\partial V_{\vec{\sigma}}} \left[h * (t_i - \hat{t}_i) * u_i\right](\vec{x}, t)\mathrm{d}\vec{x} + \int_{\partial V_{\vec{u}}} \left[h * t_i * \hat{u}_i\right](\vec{x}, t)\mathrm{d}\vec{x} \\
& + \int_{\partial V_{\varphi}} \left[h * d * \hat{\varphi}\right](\vec{x}, t)\mathrm{d}\vec{x} + \int_{\partial V_{\vec{D}}} \left[h * (d - \hat{d}) * \varphi\right](\vec{x}, t)\mathrm{d}\vec{x}\,.
\end{aligned}
\tag{1.53}
$$

Then, S is a solution of the initial mixed-boundary value problem if and only if $\delta\vartheta_t(S) = 0$ over Ω, within the time interval $t \in [0, \infty)$.

Next, let us write a specific variational form for an admissible piezoelectric process satisfying the strain-displacement relation:

Theorem 1.7 *Let Ω denote the set of all admissible processes which satisfy (1.32). Let $S = [\vec{u}, \vec{\varepsilon}, \vec{\sigma}, \vec{E}, \vec{D}, \varphi]$ be an element of Ω and define the functional Ξ_t on Ω at each time, say $t \in [0, \infty)$, by*

$$
\begin{aligned}
\Xi_t(S) =& \int_V \left[h * \sigma_{ij} * \varepsilon_{ij}\right](\vec{x}, t)\mathrm{d}\vec{x} - \frac{1}{2}\int_V S_{ijkl}(\vec{x})\left[h * \sigma_{ij} * \sigma_{kl}\right](\vec{x}, t)\mathrm{d}\vec{x} \\
& - \int_V d^E_{kij}(\vec{x})\left[h * \sigma_{ij} * E_k\right](\vec{x}, t)\mathrm{d}\vec{x} - \frac{1}{2}\int_V \chi^E_{ij}(\vec{x})\left[h * E_i * E_j\right](\vec{x}, t)\mathrm{d}\vec{x} \\
& + \int_V [h * D_i * E_i](\vec{x}, t)\mathrm{d}\vec{x} - \int_V [h * (D_{i,i} - q_e) * \varphi](\vec{x}, t)\mathrm{d}\vec{x} \\
& + \frac{1}{2}\int_V \rho(\vec{x})\,[u_i * u_i]\,(\vec{x}, t)\mathrm{d}\vec{x} - \int_V \left[f^b_i * u_i\right](\vec{x}, t)\mathrm{d}\vec{x} \\
& - \int_{\partial V_{\vec{u}}} \left[h * t_i * (u_i - \hat{u}_i)\right](\vec{x}, t)\mathrm{d}\vec{x} - \int_{\partial V_{\vec{\sigma}}} \left[h * \hat{t}_i * u_i\right](\vec{x}, t)\mathrm{d}\vec{x} \\
& + \int_{\partial V_{\varphi}} \left[h * d * \hat{\varphi}\right](\vec{x}, t)\mathrm{d}\vec{x} + \int_{\partial V_{\vec{D}}} \left[h * (d - \hat{d}) * \varphi\right](\vec{x}, t)\mathrm{d}\vec{x}\,.
\end{aligned}
\tag{1.54}
$$

Then, S is a solution of the initial mixed-boundary value problem if and only if $\delta\, \Xi_t(S) = 0$ over Ω, within the time interval $t \in [0, \infty)$.

Definition 1.3 An admissible piezoelectric process is called a *kinematically admissible piezoelectric process* if it satisfies the kinematic equations (1.31) and (1.32), the constitutive equations (1.36), and the essential boundary conditions, i.e., $(1.34)_1$ and $(1.34)_3$.

Consequently, the following variational form corresponds to a kinematically admissible piezoelectric process:

Theorem 1.8 *Let Ω denote the set of all kinematically admissible processes. Let $S = [\vec{u}, \vec{\varepsilon}, \vec{\sigma}, \vec{E}, \vec{D}, \varphi]$ be an element of Ω and define the functional $\Sigma_t(S)$ on Ω at each time, say $t \in [0, \infty)$, by*

$$
\begin{aligned}
\Sigma_t(S) =& \frac{1}{2} \int_V \left[h * \sigma_{ij} * \varepsilon_{ij}\right](\vec{x}, t)\, \mathrm{d}\vec{x} + \frac{1}{2} \int_V \rho(\vec{x})\, [u_i * u_i]\, (\vec{x}, t)\, \mathrm{d}\vec{x} \\
& - \int_V \left[f_i^b * u_i\right](\vec{x}, t)\, \mathrm{d}\vec{x} - \frac{1}{2} \int_V [h * D_i * E_i]\, (\vec{x}, t) \mathrm{d}\vec{x} \\
& + \int_V [h * q_e * \varphi]\, (\vec{x}, t) \mathrm{d}\vec{x} - \int_{\partial V_{\vec{\sigma}}} \left[h * \hat{t}_i * u_i\right](\vec{x}, t) \mathrm{d}\vec{x} \\
& - \int_{\partial V_{\vec{D}}} \left[h * \hat{d} * \varphi\right](\vec{x}, t) \mathrm{d}\vec{x}\,.
\end{aligned}
\tag{1.55}
$$

Then, S is a solution of the initial mixed-boundary value problem if and only if $\delta\, \Sigma_t(S) = 0$ over Ω, within the time interval $t \in [0, \infty)$.

Consequently, one can obtain a variational form in terms of the displacement and electric potential. To this end, the definitions of an admissible displacement-potential process and a kinematically admissible displacement-potential are required.

Definition 1.4 An array $S = [\vec{u}, \varphi]$ is called an *admissible displacement-potential process* if $\vec{u} \in C^{1,2}$, $\varphi \in C^{1,0}$.

Definition 1.5 An array $S = [\vec{u}, \varphi]$ is called a *kinematically admissible displacement-potential process* if it is an admissible displacement-potential process satisfying $(1.34)_1$ and $(1.34)_3$.

Theorem 1.9 *Let Ω denote the set of all kinematically admissible displacement-potential processes. Let $S = [\vec{u}, \varphi]$ be an element of Ω and define the functional Θ_t on Ω at each time, say $t \in [0, \infty)$, by*

$$
\begin{aligned}
\Theta_t(S) =& \frac{1}{2}\int\limits_V \left[h * (C_{ijkl}u_{k,l} + e^E_{kij}\varphi_{,k}) * u_{i,j}\right](\vec{x},t)\mathrm{d}\vec{x} \\
&+ \frac{1}{2}\int\limits_V \left[h * (e^E_{ikl}u_{k,l} - \kappa^E_{ij}\varphi_{,j}) * \varphi_{,i}\right](\vec{x},t)\mathrm{d}\vec{x} - \int\limits_V \left[f^b_i * u_i\right](\vec{x},t)\mathrm{d}\vec{x} \\
&+ \int\limits_V [h * q_e * \varphi](\vec{x},t)\mathrm{d}\vec{x} + \frac{1}{2}\int\limits_V \rho(\vec{x})\,[u_i * u_i](\vec{x},t)\,\mathrm{d}\vec{x} \\
&- \int\limits_{\partial V_{\vec{\sigma}}} \left[h * \hat{t}_i * u_i\right](\vec{x},t)\,\mathrm{d}\vec{x} - \int\limits_{\partial V_{\vec{D}}} \left[h * \hat{d} * \varphi\right](\vec{x},t)\mathrm{d}\vec{x}\,.
\end{aligned}
\tag{1.56}
$$

Then, $S = [\vec{u}, \varphi]$ is a solution of the initial mixed-boundary value problem if and only if $\delta\,\Theta_t(S) = 0$ over Ω, within the time interval $t \in [0, \infty)$.

When mechanical boundary conditions are traction-type, it is more convenient to form variational principles in terms of stress and an electric potential, for which the exact satisfaction of those conditions are guaranteed. With this in mind, we first establish a functional in terms of stress and an electric potential function for mixed mechanical boundary conditions.

Definition 1.6 An array $[\vec{\sigma}, \varphi]$ in which $\vec{\sigma}$ is a second-order symmetric tensor and $\vec{\sigma} \in C^{2,0}$, $\varphi \in C^{2,0}$ is called an *admissible stress piezoelectric process* if $\varphi = \hat{\varphi}(\vec{x}, t)$ on $\partial V_\varphi \times [0, \infty)$.

Now, we have:

Theorem 1.10 *Let Ω denote the set of all admissible stress piezoelectric processes. Let $S = [\vec{\sigma}, \varphi]$ be an element of Ω and define the functional Υ_t on Ω at each time, say $t \in [0, \infty)$, by*

$$
\begin{aligned}
\Upsilon_t(S) =& \frac{1}{2}\int\limits_V \left[\frac{h}{\rho} * \sigma_{ij,j} * \sigma_{ik,k}\right](\vec{x},t)\mathrm{d}\vec{x} - \int\limits_V \left[\left(\frac{1}{\rho}f^b_i\right)_{,j} * \sigma_{ij}\right](\vec{x},t)\mathrm{d}\vec{x} \\
&+ \frac{1}{2}\int\limits_V [S_{ijkl}\sigma_{ij} * \sigma_{kl} + \chi^E_{ij}\varphi_{,i} * \varphi_{,j}](\vec{x},t)\mathrm{d}\vec{x} \\
&+ \int\limits_V [-d^E_{ikl}\sigma_{kl} * \varphi_{,i} - q_e * \varphi](\vec{x},t)\mathrm{d}\vec{x} + \int\limits_{\partial V_{\vec{u}}} \left[\left(\frac{f^b_i}{\rho} - \hat{u}_i\right) * t_i\right](\vec{x},t)\mathrm{d}\vec{x} \\
&+ \int\limits_{\partial V_{\vec{\sigma}}} \left[\frac{h}{\rho} * (\hat{t}_i - t_i) * \sigma_{ij,j}\right](\vec{x},t)\mathrm{d}\vec{x} + \int\limits_{\partial V_{\vec{D}}} \left[\hat{d} * \varphi\right](\vec{x},t)\mathrm{d}\vec{x}\,.
\end{aligned}
\tag{1.57}
$$

Then, S is a solution of the initial mixed-boundary value problem if and only if $\delta\,\Upsilon_t(S) = 0$ over Ω, within the time interval $t \in [0, \infty)$.

Definition 1.7 An admissible stress piezoelectric process is called a *dynamically admissible stress piezoelectric process* if $\sigma_{ij}n_j = \hat{t}_i(\vec{x}, t)$ on $\partial V_{\vec{\sigma}} \times [0, \infty)$.

Starting from $\Upsilon_t(S)$, one may arrive at a variational form for a dynamically admissible stress piezoelectric process:

Theorem 1.11 *Let Ω be the set of all dynamically admissible stress piezoelectric processes. Let $S = [\vec{\sigma}, \varphi]$ be an element of Ω and define the functional $\Im_t$ on Ω at each time, say $t \in [0, \infty)$, by*

$$\begin{aligned}\Im_t(S) =& \frac{1}{2}\int_V \left[\frac{h}{\rho} * \sigma_{ij,j} * \sigma_{ik,k}\right](\vec{x}, t)\mathrm{d}\vec{x} - \int_V \left[\left(\frac{1}{\rho} f_i^b\right)_{,j} * \sigma_{ij}\right](\vec{x}, t)\mathrm{d}\vec{x} \\ &+ \frac{1}{2}\int_V \left[S_{ijkl}\sigma_{ij} * \sigma_{kl} + \chi_{ij}^E \varphi_{,i} * \varphi_{,j}\right](\vec{x}, t)\mathrm{d}\vec{x} \\ &- \int_V \left[d_{ikl}^E \sigma_{kl} * \varphi_{,i}\right](\vec{x}, t)\mathrm{d}\vec{x} - \int_V [q_e * \varphi](\vec{x}, t)\mathrm{d}\vec{x} + \int_{\partial V_{\vec{D}}} \left[\hat{d} * \varphi\right](\vec{x}, t)\mathrm{d}\vec{x}\,. \end{aligned} \tag{1.58}$$

Then, S is a solution of the traction problem (i.e., $\partial V_{\vec{u}} = \varnothing$) if and only if $\delta\, \Im_t(S) = 0$ over Ω within the time interval $t \in [0, \infty)$.

1.5 Random Piezoelectric Fields

1.5.1 Basic Considerations

This section provides a bridge to subsequent chapters on random fields. The dependent fields are stochastic because they depend on space and time, while the material property fields only depend on space, so they are just tensor-valued random fields (TRFs). First, given the fact that no material continuum is perfectly homogeneous and that mass density is one of the basic continuum properties, it should be taken as a random field (RF)

$$\left\{\rho\left(\vec{x}, \omega\right); \vec{x} \in \mathbb{R}^3, \omega \in \Omega\right\}. \tag{1.59}$$

Here $\vec{x}$ is the location in the material domain while ω indicates one realization of the mass density field ρ from a probability space. The mass density is defined through a straightforward integration over the mesoscale domain:

$$\rho_\delta\left(\vec{x}, \omega\right) = \frac{1}{V_\delta}\int_{B_\delta} \rho\left(\vec{x}, \omega\right) \mathrm{d}V\,,$$

where V_δ is the volume of B_δ, while $\rho\left(\vec{x}, \omega\right)$ is the density at the finest measurable scale, the resolution of a given measurement technique. In the language of random

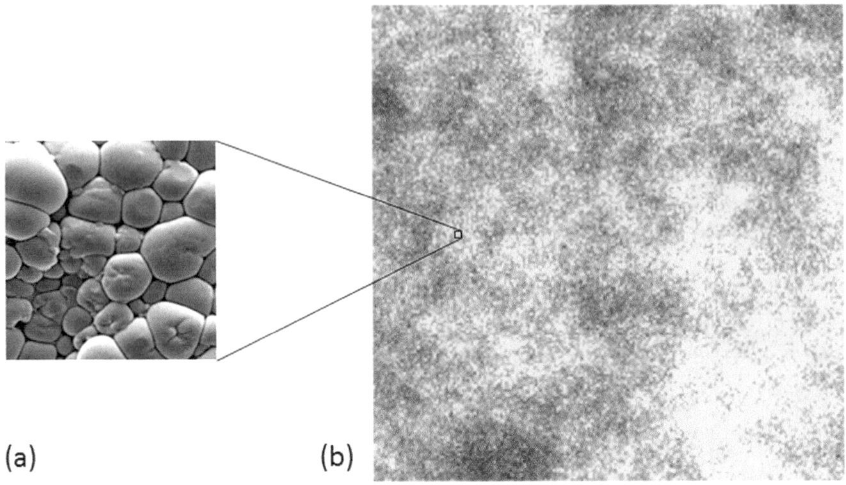

Fig. 1.1 The mesoscale domain B_δ of a disordered polycrystal (**a**) is the statistical volume element (SVE) of the random field. *Source* https://www.bing.com/images (**b**) Note that, the larger the SVE, the smoother the resulting random field

processes, this is akin to the so-called *local averaging* of random fields. The situation is illustrated in Fig. 1.1: mass density, being a continuum property, depends on the volume under consideration.

The mesoscale domain B_δ of lengthscale L, for a microstructure of a typical grain size d, is parametrised by a dimensionless *mesoscale*

$$\delta := \frac{L}{d}.$$

By analogy to (1.59), considering a linear elastic response of B_δ , the mechanical properties are described by an RF of the in-plane stiffness tensor

$$\left\{ \mathsf{C}_\delta \left(\vec{x}, \omega \right); \vec{x} \in \mathbb{R}^2, \omega \in \Omega \right\}.$$

The same goes for the electrical permittivity $\vec{\kappa}$ and the piezoelectric property $\vec{e}^E$. Thus, we have a four-component RF

$$\left\{ \rho_\delta \left(\vec{x}, \omega \right), \mathsf{C}_\delta \left(\vec{x}, \omega \right), \vec{\kappa}_\delta^E \left(\vec{x}, \omega \right), \vec{e}_\delta^E \left(\vec{x}, \omega \right); \vec{x} \in \mathbb{R}^3, \omega \in \Omega \right\} \tag{1.60}$$

in which the components are rank 0, rank 4, rank 2, and rank 3 tensor random fields. By mathematical analogy, the situation is identical in the case of piezomagnetic properties.

This set of all the deterministic realisations defines a random material $\mathcal{B}_\delta$ parametrised by a mesoscale δ, and occupying a domain $B_\delta \subset \mathbb{E}^2$:

$$\mathcal{B}_\delta = \{B_\delta(\omega); \omega \in \Omega\} .$$

Essentially, the finest resolution $L \to 0$ in (1.60) corresponds to $\delta \to 0$. With coarsening observation, the mesoscale δ increases and the randomness of properties decreases: the continuum RF model (1.60) is mesoscale-dependent.

In the limit $\delta \to \infty$, the randomness is expected to die to zero, and the material property fields should become deterministic. For example, take

$$\lim_{\delta\to\infty} \rho(\vec{x}, \omega, \delta) = \rho^{eff}, \quad \lim_{\delta\to\infty} \mathsf{C}_\delta(\vec{x}, \omega) = \mathsf{C}^{eff} , \tag{1.61}$$

where, provided the RF is spatially homogeneous (a concept to be made rigorous in the next chapters), a conventional, deterministic piezoelectric continuum is obtained. In the above, we have introduced the effective material properties (such as stiffness, modulus, and Poisson ratio) which are typically employed in deterministic models of continuum mechanics.

- How can one determine the SVE (mesoscale) properties?
- How can one solve a macroscopic boundary value problem?
- How can one proceed in the case of media with fractal structures and (also) long range effects?

Note that the stiffness tensor cannot simply be locally averaged like the mass density in (1.59), lest we would obtain a Voigt-type (very stiff) estimate; averaging the compliance tensor would result in a Reuss-type (very soft) average. Overall, the mesoscale tensorial properties require a procedure eventually upscaling the mesoscale properties towards the effective ones such as C^{eff} in the second equation in (1.61).

1.5.2 *Determination of Mesoscale Properties Using a Homogenization Condition*

In order to carry out the mesoscale upscaling and smoothing, we employ a scale-dependent homogenization in the vein of the Hill–Mandel condition, e.g. [26]. The objective is to replace a heterogeneous linear elastic medium by a homogeneous linear elastic one. Since the mechanical work going into a volume V_δ is $W = \frac{V_\delta}{2}\sigma_{ij}\varepsilon_{ij}$, the said condition requires that the volume average of the scalar product of the stress and strain fields equals the scalar product of their volume averages

$$\begin{array}{ccc} \frac{1}{2}\,\overline{\sigma_{ij}\,\varepsilon_{ij}} & = & \frac{1}{2}\,\overline{\sigma_{ij}}\;\overline{\varepsilon_{ij}} \\ \textbf{energetic interpretation} & = & \textbf{mechanical interpretation,} \end{array} \tag{1.62}$$

where, for a function $f(\omega, \vec{x})$, the overbar signifies the volume average over the mesoscale domain $B_\delta(\omega)$, i.e. $\overline{f(\omega)} = \frac{1}{V}\int_{B_\delta(\omega)} f(\omega, \vec{x})\mathrm{d}V$. In other words, the left and right-hand sides of (1.29) express two different ways of looking at the work going into the material per unit of its volume.

Turning to the piezoelectricity, we start from a more general expression

$$W = \frac{1}{2}\sigma_{ij}\varepsilon_{ij} - \frac{1}{2}E_i D_i ,$$

which, upon substitution of the constitutive relations (1.36), i.e.

$$\sigma_{ij} = C_{ijkl}\varepsilon_{kl} - e^E_{ijk}E_k \; D_i = e^E_{ijk}\varepsilon_{kl} + \kappa^E_{ij}E_j ,$$

becomes the total free energy (a so-called electric enthalpy)

$$H\left(\vec{\varepsilon}, \vec{E}\right) = \frac{1}{2}\varepsilon_{ij}C_{ijkl}\varepsilon_{kl} - E_k e^E_{ijk}\varepsilon_{ij} - \frac{1}{2}E_i\kappa^E_{ij}E_j .$$

At this point, recalling the inverse constitutive relations (1.37), we observe that the direct piezoelectric tensor is $d^E_{ijp} = S_{ijkl}e^E_{klp}$, while the reverse piezoelectric tensor is $e^E_{klp} = -C_{klij}d^E_{ijp}$.

By analogy to (1.62), there are two ways of volume averaging W:

$$\begin{array}{ccc} \frac{1}{2}\,\overline{\sigma_{ij}\,\varepsilon_{ij}} - \frac{1}{2}\,\overline{E_i\,D_i} & = & \frac{1}{2}\,\overline{\sigma_{ij}}\;\overline{\varepsilon_{ij}} - \frac{1}{2}\,\overline{E_i}\;\overline{D_i} \\ \textbf{energetic interpretation} & = & \textbf{mechanical interpretation.} \end{array} \tag{1.63}$$

Now, consider the stress (σ_{ij}) and strain (ε_{ij}) fields as superpositions of their means ($\overline{\sigma_{ij}}$ and $\overline{\varepsilon_{ij}}$) with zero-mean fluctuations (σ'_{ij} and ε'_{ij})

$$\sigma_{ij}(\omega, \vec{x}) = \overline{\sigma_{ij}} + \sigma'_{ij}(\omega, \vec{x}), \quad \varepsilon_{ij}(\omega, \vec{x}) = \overline{\varepsilon_{ij}} + \varepsilon'_{ij}(\omega, \vec{x}) .$$

Similarly, for the electrical (E_i) and electrical displacement (D_i) fields:

$$E_i(\omega, \vec{x}) = \overline{E_i} + E'_i(\omega, \vec{x}), \quad D_i(\omega, \vec{x}) = \overline{D_i} + D'_i(\omega, \vec{x}) .$$

Substituting these into (1.63), we find for the volume average of the energy density over $B_\delta(\omega)$, we obtain two conditions for volume-type cross-correlations

$$\overline{\sigma'_{ij}\,\varepsilon'_{ij}} = 0, \quad \overline{E'_i\,D'_i} = 0 .$$

This implies a spatial uncorrelatedness of fluctuations of stress and strain fields on one hand, and of electrical and electrical displacement on the other. These conditions can be satisfied by any one of three different types of uniform boundary conditions (BCs) for boundary value problems on the mesoscale domain.

1. *Uniform Dirichlet BC*

$$u_i(\vec{x}) = \varepsilon^0_{ij}\, x_j \quad \forall \vec{x} \in \partial B_\delta\,,$$

and

$$\phi(\vec{x}) = \phi^0{,_j}\; x_j \quad \forall \vec{x} \in \partial B_\delta\,.$$

In the above, ε^0_{ij} is the constant strain, while $\phi^0{,_j}$ is the constant gradient of a scalar electric potential ϕ. [Note that, in general, $E_j = -\phi,_j$.]

2. *Uniform Neumann BC*

$$t_i^{(n)}(\vec{x}) = \sigma^0_{ij}\, n_j \quad \forall \vec{x} \in \partial B_\delta\,, \tag{1.64}$$

and

$$D(\vec{x}) = D^{(n)0} \quad \forall \vec{x} \in \partial B_\delta\,.$$

In the above, σ^0_{ij} is the constant strain, while $D^{(n)0}$ is a constant vector.

3. *Uniform mixed-orthogonal BC*

$$\left[u_i(\vec{x}) - \varepsilon^0_{ij}\, x_j\right]\left[t_i^{(n)}(\vec{x}) - \sigma^0_{ij}\, n_j\right] = 0 \quad \forall \vec{x} \in \partial B_\delta\,.$$

With reference to [17, 26], it follows from the variational principles that the Dirichlet (respectively, Neumann) type BCs provide upper (lower) estimates on the mechanical and electrical responses of the mesoscale domain $B_\delta\,(\omega)$. More specifically, setting $E_i = D_i = 0$ results in a purely mechanical problem for $C^{(\vec{\varepsilon})}_{klij}\,(\delta,\omega)$ under the Dirichlet BC, and for $S^{(\vec{\sigma})}_{klij(\delta)}$ under the Neumann BC. On the other hand, setting $\varepsilon_{ij} = \sigma_{ij} = 0$ results in a purely electrical problem for $\kappa^{E(\vec{\varepsilon})}_{ij(\delta)}$ under the Dirichlet BC, and for $\chi^{E(\vec{\varepsilon})}_{ij(\delta)}$ under the Neumann BC.

As always, the explicit dependence on δ indicates that these properties are mesoscale-dependent, while the dependence on ω that they are random. Upon computing, in a Monte Carlo sense, a number of stochastic boundary value problems for different realisations of the random mesoscale domains $\{B_\delta\,(\omega)\,;\,\omega \in \Omega\}$, one can assess the statistics of random tensors

$$C^{(\vec{\varepsilon})}_{klij(\delta)}\,(\omega)\,,\; S^{(\vec{\sigma})}_{klij(\delta)}\,(\omega)\,,\; \kappa^{E(\vec{\varepsilon})}_{ij(\delta)}\,(\omega)\,,\; \chi^{E(\vec{\varepsilon})}_{ij(\delta)}\,(\omega)\;.$$

In the first place, upon statistical averaging, one obtains mesoscale bounds on the macroscale (effective) properties C^{eff}_{klij} and $\kappa^{E,eff}_{ij}$:

$$\begin{aligned}
\left\langle S^{(\vec{\sigma})}_{klij}\,(\delta)\right\rangle^{-1} &\leq C^{eff}_{klij} \leq \left\langle C^{(\vec{\varepsilon})}_{klij}\,(\delta)\right\rangle, \\
\left\langle \chi^{E(\vec{\varepsilon})}_{ij}\,(\delta)\right\rangle^{-1} &\leq \kappa^{E,eff}_{ij} \leq \left\langle \kappa^{E(\vec{\varepsilon})}_{ij}\,(\delta)\right\rangle.
\end{aligned}$$

Increasing (respectively, decreasing) the mesoscale domain size and repeating the procedure would result in tighter (wider) bounds on C^{eff}_{klij} and $\kappa^{E,eff}_{ij}$:

$$\begin{aligned} \left\langle S^{(\vec{\sigma})}_{klij(\delta')}\right\rangle^{-1} \leq \left\langle S^{(\vec{\sigma})}_{klij(\delta)}\right\rangle^{-1} \leq C^{eff}_{klij} \leq \left\langle C^{(\vec{\varepsilon})}_{klij(\delta)}\right\rangle \leq \left\langle C^{(\vec{\varepsilon})}_{klij(\delta')}\right\rangle, \quad \delta' < \delta, \\ \left\langle \chi^{E(\vec{\varepsilon})}_{ij(\delta')}\right\rangle^{-1} \leq \left\langle \chi^{E(\vec{\varepsilon})}_{ij(\delta)}\right\rangle^{-1} \leq \kappa^{E,eff}_{ij} \leq \left\langle \kappa^{E(\vec{\varepsilon})}_{ij(\delta)}\right\rangle \leq \left\langle \kappa^{E(\vec{\varepsilon})}_{ij(\delta')}\right\rangle, \quad \delta' < \delta. \end{aligned} \tag{1.65}$$

These are hierarchies of mesoscale bounds describing the scaling from SVE towards the RVE of deterministic continuum physics. The trend towards the RVE is compactly described in terms of the so-called scaling functions [29, 30].

Note that the mixed-orthogonal BCs (1.36) yield intermediate responses (i.e., between the Dirichlet-type and Neumann-type bounds), which do not display clear scaling.

In order to obtain the mesoscale coupled piezoelectric responses, we envisage two loading programs:

Finding $d^E_{ijp(\delta)}$:

(i) apply a stress-free BC (1.64) with a nonzero electric field $E_j = -\phi^0{}_{,j}$;
(ii) calculate the resulting volume averaged strain $\overline{\varepsilon_{ij}}$;
(iii) calculate $d^E_{ijp(\delta)} = \left(S^{(\vec{\sigma})}_{ijkl(\delta)}\right)^{-1} \overline{\varepsilon_{kl}}$.

Finding $e^E_{ijp(\delta)}$:

(i) apply a zero-strain BC (1.33) with a nonzero electric displacement field D_j;
(ii) calculate the resulting volume averaged stress $\overline{\sigma_{ij}}$;
(iii) calculate $e^E_{ijp(\delta)} = \left(C^{(\vec{\varepsilon})}_{klij(\delta)}\right)^{-1} \overline{\sigma_{ij}}$.

Next, carrying out these loading programs in the Monte Carlo sense, taking ensemble averages, and noting the scale-dependent hierarchies (1.65), one should be able to establish hierarchies of mesoscale bounds on the macroscale (effective) $\left\langle d^{E,eff}_{ijp}\right\rangle$ and $\left\langle e^{E,eff}_{ijp}\right\rangle$, similar to those established for the thermal expansion coefficients in [18]. This research is presently being planned.

Chapter 2
Mathematical Preliminaries

2.1 Piezoelectricity Classes

How many different classes of piezoelectric materials exist? To answer this question, we need a little bit more mathematics.

Let O(3) be the group of all orthogonal 3×3 matrices. Let P be the linear space of all piezoelectric tensors d^E_{ijk} with inner product

$$(\mathsf{d}^E_{ijk}, \mathsf{f}^E_{ijk}) = \mathsf{d}^E_{ijk}\mathsf{f}^E_{ijk}\,.$$

The group O(3) acts on P by

$$(g\mathsf{d}^E)_{ijk} = g_{il}g_{jm}g_{kn}e_{lmn}, \qquad g \in \mathrm{O}(3), \quad \mathsf{d}^E \in \mathsf{P}\,. \tag{2.1}$$

For a given piezoelectric tensor d^E, its *orbit* is defined as

$$O_{\mathsf{d}^E} = \{\, g\mathsf{d}^E : g \in \mathrm{O}(3)\,\}\,.$$

The set G_{d^E} of all $g \in \mathrm{O}(3)$ with $g\mathsf{d}^E = \mathsf{d}^E$ is a closed subgroup of the group O(3) called the *stationary subgroup* of d^E. It is easy to see that the stationary subgroup of a point $g\mathsf{d}^E \in O_{\mathsf{d}^E}$ is

$$G_{g\mathsf{d}^E} = \{\, g\mathsf{h}g^{-1} : \mathsf{h} \in G_{\mathsf{d}^E}\,\}\,,$$

that is, a group *conjugate* to G_{d^E}. As g runs over the group O(3), the point $g\mathsf{d}^E$ runs over the orbit O_{d^E}, and the group $G_{g\mathsf{d}^E}$ runs over the *conjugacy class* $[G_{\mathsf{d}^E}]$ of the group G_{d^E}, that is, over the set of all groups conjugate to G_{d^E}.

We constructed a map from the set of all orbits of the group O(3) to the set of conjugacy classes of its closed subgroups. Mathematically, the conjugacy classes that belong to the image of the above map are called the *piezoelectricity classes*.

A. Malyarenko et al., *Random Fields of Piezoelectricity and Piezomagnetism*, SpringerBriefs in Mathematical Methods,
https://doi.org/10.1007/978-3-030-60064-8_2

Physically, a piezoelectric material belongs to a given piezoelectricity class $[G]$ if and only if its piezoelectric tensor d^E belongs to the *fixed point set* of G:

$$\mathsf{V}_G = \{\, \mathsf{d}^E \in \mathsf{P} \colon g\mathsf{d}^E = \mathsf{d}^E \quad \text{for all} \quad g \in G \,\}.$$

It turns out that there are 16 piezoelectricity classes, see [23]. To formulate the result, we need to assign names to all conjugacy classes of closed subgroups of the group O(3).

Theorem 2.1 ([16, 24, 32]) *Every closed subgroup of* O(3) *is conjugate to one of the groups in the following list.*

1. $\{\, Z_n \colon n \geq 1 \,\}$, $\{\, D_n \colon n \geq 2 \,\}$, $\mathcal{T}$, $\mathcal{O}$, $\mathcal{I}$, SO(2), O(2), SO(3).
2. $G \times Z_2^c$, *where* G *is one of the groups listed in item 1, and* $Z_2^c = \{\delta_{ij}, -\delta_{ij}\}$.
3. $\{\, Z_{2n}^- \colon n \geq 1 \,\}$, $\{\, D_n^v \colon n \geq 2 \,\}$, $\{\, D_{2n}^h \colon n \geq 2 \,\}$, $\mathcal{O}^-$, $\mathrm{O}(2)^-$.

To explain the introduced symbols, we follow [24]. Let the vectors $\vec{e}^1$, $\vec{e}^2$, and $\vec{e}^3$ constitute the standard basis of $\mathbb{R}^3$. Denote by $Q(\vec{v}, \vartheta)$ the rotation about $\vec{v} \in \mathbb{R}^3$ by angle ϑ, and by $\sigma_{\vec{v}}$ the reflection through the plane normal to the $\vec{u}$ axis.

Z_n is the *cyclic group* of order n generated by $Q(\vec{e}^3, 2\pi/n)$. D_n is the *dihedral group* of order $2n$ generated by Z_n and $Q(\vec{e}^1, \pi)$. The *tetrahedral group* $\mathcal{T}$ of order 12 fixes a tetrahedron. The *octahedral group* $\mathcal{O}$ of order 24 fixes an octahedron or a cube. The *icosahedral group* $\mathcal{I}$ of order 60 fixes an icosahedron or a dodecahedron. SO(2) is the group of rotations $Q(\vec{e}^3, \vartheta)$ with $\vartheta \in [0, 2\pi)$. O(2) is the group generated by SO(2) and $Q(\vec{e}^1, \pi)$. SO(3) is the group of orthogonal matrices with unit determinant. Z_2^- is the order 2 *reflection group* generated by $\sigma_{\vec{e}^1}$. Z_{2n}^- with $n \geq 2$ is the group of order $2n$ generated by $Q(\vec{e}^3, \pi/n)$ and $\sigma_{\vec{e}^3}$. D_{2n}^h is the *prismatic group* of order $4n$ generated by Z_{2n}^- and $Q(\vec{e}^1, \pi)$. D_n^v is the *pyramidal group* of order $2n$ generated by Z_n and $\sigma_{\vec{e}^1}$. The group $\mathcal{O}^-$ of order 24 has the same rotation axes as $\mathcal{T}$ but with six mirror planes, each through two 3-fold axes. $\mathrm{O}(2)^-$ is the group generated by SO(2) and $\sigma_{\vec{e}^1}$. See also [24, Appendix B], where the correspondence between the above notation and the classical crystallographic one is given.

The 16 piezoelectricity classes are $[Z_1]$, $[Z_2]$, $[Z_3]$, $[D_2^v]$, $[D_3^v]$, $[Z_2^-]$, $[Z_4^-]$, $[D_2]$, $[D_3]$, $[D_4^h]$, $[D_6^h]$, [SO(2)], [O(2)], $[\mathrm{O}(2)^-]$, $[\mathcal{O}^-]$ and [O(3)].

Consider the group

$$G_{\max} = \{\, g \in \mathrm{O}(3) \colon g\mathsf{v} \in \mathsf{V}_G \quad \text{for all} \quad \mathsf{v} \in \mathsf{V}_G \,\}.$$

Obviously, $G \subset G_{\max}$. It turns out that $G_{\max}$ is the normaliser of the group G in O(3):

$$G_{\max} = \{\, h \in \mathrm{O}(3) \colon hGh^{-1} = G \,\}.$$

In what follows, we denote by K a closed subgroup of $G_{\max}$ such that G is a closed subgroup of K. Note that the action of K on V_G given by (2.1) has the following property: $g\mathsf{d}^E$ is an orthogonal linear operator on V_G. In other words, the map $g \mapsto g\mathsf{v}$ is an *orthogonal representation* of the group K in the space V_G.

From now on, following [1], we denote a representation of a compact topological group by the same symbol as the finite-dimensional linear space where it acts. If it is necessary to consider the *operators* of a generic representation, we denote them by $\theta(g)$. If the operators have a standard notation, we use it instead.

2.2 A Random Field Approach

At microscopic length scales, *spatial randomness* of the piezoelectric material needs to be taken into account. Mathematically, we consider the piezoelectric tensor $\mathsf{d}^E(\vec{x})$ as a single realisation of a *random field*. That is, for each position vector $\vec{x}$ inside a piezoelectric body $\mathcal{B} \subset \mathbb{R}^3$, there is a random tensor $\mathsf{d}^E(\vec{x})$ taking values in the space V_G.

We assume that the random field $\mathsf{d}^E(\vec{x})$ is *second-order*, that is, $\mathsf{E}[\|\mathsf{d}^E(\vec{x})\|^2] < \infty$ for all $\vec{x} \in \mathcal{B}$. Moreover, assume that the field $\mathsf{d}^E(\vec{x})$ is *mean-square continuous*, that is, for any $\vec{x}_0 \in \mathcal{B}$ we have

$$\lim_{\|\vec{x}-\vec{x}_0\|\to 0} \mathsf{E}[\|\mathsf{d}^E(\vec{x}) - \mathsf{d}^E(\vec{x}_0)\|^2] = 0\,.$$

Under a shift of the Cartesian coordinate system, for any positive integer n and for any distinct points $\vec{x}_1, \dots, \vec{x}_n \in \mathcal{B}$, the V^n_G-valued finite-dimensional distributions $(\mathsf{d}^E(\vec{x}_1), \dots, \mathsf{d}^E(\vec{x}_n))$ do not change. In particular, the *one-point correlation tensor* of the random field $\mathsf{d}^E(\vec{x})$

$$\langle \mathsf{d}^E(\vec{x})\rangle = \mathsf{E}[\mathsf{d}^E(\vec{x})]$$

does not depend on $\vec{x} \in \mathcal{B}$, while its *two-point correlation tensor*

$$\langle \mathsf{d}^E(\vec{x}), \mathsf{d}^E(\vec{y})\rangle = \mathsf{E}[(\mathsf{d}^E(\vec{x}) - \langle \mathsf{d}^E(\vec{x})\rangle) \otimes (\mathsf{d}^E(\vec{y}) - \langle \mathsf{d}^E(\vec{y})\rangle)]$$

depends only on the difference $\vec{y} - \vec{x}$. Such a field is called *wide-sense homogeneous*. It is convenient to consider a random field $\{\, \mathsf{d}^E(\vec{x}) \colon \vec{x} \in \mathcal{B}\,\}$ as the restriction to $\mathcal{B}$ of a random field defined on all of $\mathbb{R}^3$.

What happens when one applies an orthogonal transformation $g \in K$ to the Cartesian coordinate system? A position vector $\vec{x} \in \mathbb{R}^3$ becomes the vector $g\vec{x}$. The random tensor $\mathsf{d}^E(\vec{x})$ becomes the tensor $\mathsf{d}^E(g\vec{x}) = g\mathsf{d}^E(\vec{x})$. The one- and two-point correlation tensors of the transformed field must be equal to those of the original field:

$$\begin{aligned}\langle \mathsf{d}^E(g\vec{x})\rangle &= g\langle \mathsf{d}^E(\vec{x})\rangle\,,\\ \langle \mathsf{d}^E(g\vec{x}), \mathsf{d}^E(g\vec{y})\rangle &= (g\otimes g)\langle \mathsf{d}^E(\vec{x}), \mathsf{d}^E(\vec{y})\rangle\,.\end{aligned} \tag{2.2}$$

A random field satisfying these equations is called *wide-sense K-isotropic*, (or (K,θ)-isotropic, if necessary). In what follows, we will omit the words "wide-sense" and call such fields homogeneous and K-isotropic.

In this book, we calculate the correlation structure of homogeneous and K-isotropic random fields, that is, find the general form of their one- and two-point correlation tensors. Moreover, we find the spectral expansions of such fields in terms of stochastic integrals with respect to certain random measures.

It turns out that the answer depends on the choice of a basis in the space V_G. In the following chapter, we construct suitable bases in the above space.

Chapter 3
The Choice of a Basis in the Space V_G

3.1 Introduction

The space V_G carries the orthogonal representation of the group K. A linear subspace $\mathsf{W} \subseteq \mathsf{V}_G$ is called *invariant* if $g\vec{w} \in \mathsf{W}$ for all $g \in K$ and for all $\vec{w} \in \mathsf{W}$. Any orthogonal representation acting on V_G has at least two invariant subspaces: $\{\mathsf{0}\}$ and V_G. A representation is called *irreducible* if no other invariant subspaces exist.

Let V_1 and V_2 be two real finite-dimensional orthogonal representations of a group K. A linear operator $A\colon \mathsf{V}_1 \to \mathsf{V}_2$ is called *intertwining* if it commutes with the above representations:

$$A(g\mathsf{v}) = g(A\mathsf{v}), \qquad g \in K, \quad \mathsf{v} \in \mathsf{V}_1 .$$

The set of all intertwining operators is a real linear space. The representations are called *equivalent* if the above space contains at least one invertible operator.

It is well known that any real finite-dimensional orthogonal representation V_G of a compact topological group K is *completely reducible*. This means the following. Let V_i run over inequivalent irreducible orthogonal representations of K as i runs over some set of indices I. There exist unique nonnegative integers m_i of which all but a finite number are zero such that V_G is equivalent to the direct sum of m_i copies of V_i.

We choose a basis in the space V_G in such a way that for any i with $1 \le i \le N$, there is a subset of the constructed basis which is a basis of the space V_i. In the remaining part of this chapter, we explain the above construction, using examples.

A. Malyarenko et al., *Random Fields of Piezoelectricity and Piezomagnetism*,
SpringerBriefs in Mathematical Methods,
https://doi.org/10.1007/978-3-030-60064-8_3

3.2 Example: The Piezoelectricity Class $[Z_1]$

First, put $K = Z_1$. We use information about conjugacy classes of closed subgroups of the group O(3) from [2]. The authors of this book use the Schoenflies notation. According to [24, Appendix B], the Schoenflies notation for Z_1 is C_1. The group C_1 is described in [2, Table 1]. In particular, all irreducible orthogonal representations of the group K are equivalent to the trivial representation $A(E) = 1$, where $E = \delta_{ij}$ is the unique element of K. Here and below we use the notation for the group elements and group representations from [2].

It follows that the representation θ is the direct sum of 18 copies of the trivial representation A of the group K. Moreover, any one-dimensional subspace of the space V_G carries the trivial presentation of K. It follows that the basis of V_G can be chosen arbitrarily.

We choose the standard basis described in [28]. For simplicity, in what follows we omit the upper index E in the notation for the piezoelectricity tensor. Specifically, the piezoelectric tensor is symmetric with respect to its second and third indices: $d_{kij} = d_{kji}$. It follows that $\mathsf{P} = \mathbb{R}^3 \otimes \mathsf{S}^2(\mathbb{R}^3)$, where $\mathsf{S}^2(\mathbb{R}^3)$ is the 6-dimensional real linear space of all symmetric 3×3 matrices with real entries. On the other hand, by the second equation in (1.28), the piezoelectric tensor is a linear map acting from $\mathsf{S}^2(\mathbb{R}^3)$ to $\mathbb{R}^3$. Define an orthogonal operator $A\colon \mathsf{P} \to \mathbb{R}^3 \otimes \mathbb{R}^6$ as follows: A maps the 9 basis tensors $\vec{e}^i \otimes \vec{e}^j \otimes \vec{e}^j$ of the space P to the basis tensors $\vec{e}^i \otimes \vec{f}^j$ of the space $\mathbb{R}^3 \otimes \mathbb{R}^6$, the 3 basis tensors $\frac{1}{\sqrt{2}}\vec{e}^i \otimes (\vec{e}^2 \otimes \vec{e}^3 + \vec{e}^3 \otimes \vec{e}^2)$ to the basis tensors $\vec{e}^i \otimes \vec{f}^4$, the 3 basis tensors $\frac{1}{\sqrt{2}}\vec{e}^i \otimes (\vec{e}^1 \otimes \vec{e}^3 + \vec{e}^3 \otimes \vec{e}^1)$ to the basis tensors $\vec{e}^i \otimes \vec{f}^5$, and the 3 basis tensors $\frac{1}{\sqrt{2}}\vec{e}^i \otimes (\vec{e}^1 \otimes \vec{e}^2 + \vec{e}^2 \otimes \vec{e}^1)$ to the basis tensors $\vec{e}^i \otimes \vec{f}^6$. In the above described *compressed matrix notation*, a piezoelectric tensor is described by a 3×6 matrix with real entries. In what follows, we calculate the elements of various bases in the above described basis.

The normaliser of the group Z_1 is $K = \mathrm{O}(3)$. Its irreducible orthogonal representations are described as follows.

The representation H_ℓ is the linear space of real-valued homogeneous harmonic (with vanishing Laplacian) polynomials of degree $\ell \geq 0$ in three variables x, y, z of dimension $2\ell + 1$. The group K acts by

$$gp(\vec{x}) = p(g^{-1}\vec{x})\,.$$

The representation H_ℓ^* is the same linear space but with different action:

$$gp(\vec{x}) = \det g p(g^{-1}\vec{x})\,,$$

see [10, 23].

Note that the elements of the space H_ℓ are even (resp. odd) functions of the variable $\vec{x}$ whenever the number ℓ is even (resp. odd). It follows that the representations H_ℓ

with even ℓ and H_ℓ^* with odd ℓ map the matrix $-\delta_{ij}$ to the identity operator I. The representations H_ℓ with odd ℓ and H_ℓ^* with even ℓ map the matrix $-\delta_{ij}$ to the operator $-I$.

In particular, the representation that maps $g \in \mathrm{O}(3)$ to itself is equivalent to H_1. The elements of the space $\mathbb{R}^3$ where this representation acts are *pseudo-vectors* (they change sign under reflection). In the irreducible components of the representation $H_1 \otimes H_1$, the matrix $-\delta_{ij}$ maps to I. The Clebsch–Gordan formula [10, 23] gives

$$H_1 \otimes H_1 = H_0 \oplus H_1^* \oplus H_2 .$$

The one-dimensional irreducible component H_0 acts on the space of *scalars* generated by the symmetric identity matrix δ_{ij}. There are two matrices of norm 1 in this space: $\frac{1}{\sqrt{3}}\delta_{ij}$ and $-\frac{1}{\sqrt{3}}\delta_{ij}$. We choose the first matrix as one of the basis tensors of the space $\mathsf{S}^2(\mathbb{R}^3)$.

The three-dimensional irreducible component H_1^* acts on the space of 3×3 skew-symmetric matrices. The elements of the above space are vectors.

Finally, the five-dimensional irreducible component H_2 acts on the space of traceless 3×3 symmetric matrices. The elements of this space are called *deviators*. The basis in the space of deviators (and, more generally, in the space where the representation $H_{\ell_1} \otimes H_{\ell_2}$ acts) was constructed in [9]. The matrices of this basis are denoted by $g^k_{\ell[\ell_1,\ell_2]}$, where $|\ell_1 - \ell_2| \leq \ell \leq \ell_1 + \ell_2$ and $-\ell \leq k \leq \ell$. We call them the *Godunov–Gordienko matrices*. In particular, the (i, j)th entry of the matrix $g^0_{0[1,1]}$ is $g^{0[i,j]}_{0[1,1]} = \frac{1}{\sqrt{3}}\delta_{ij}$, $-1 \leq i, j \leq 1$, exactly what we chose above.

In general, the Godunov–Gordienko matrices can be calculated by the algorithm proposed in [31]. The nonzero entries of the five symmetric matrices $g^k_{2[1,1]}$, $-2 \leq k \leq 2$, that are located on and over the main diagonal, are as follows:

$$\begin{aligned} g^{-2[-1,1]}_{2[1,1]} &= g^{2[1,1]}_{2[1,1]} = -\frac{1}{\sqrt{2}}, \\ g^{-1[-1,0]}_{2[1,1]} &= g^{1[0,1]}_{2[1,1]} = g^{2[-1,-1]}_{2[1,1]} = \frac{1}{\sqrt{2}}, \\ g^{0[-1,-1]}_{2[1,1]} &= g^{0[1,1]}_{2[1,1]} = -\frac{1}{\sqrt{6}}, \\ g^{0[0,0]}_{2[1,1]} &= \frac{\sqrt{2}}{\sqrt{3}}. \end{aligned} \tag{3.1}$$

The orthogonal representation (2.1) becomes

$$H_1 \otimes (H_0 \oplus H_2) = H_1 \oplus H_1 \oplus H_2^* \oplus H_3 .$$

Note that this representation does not contain trivial components. In other words, $\mathsf{V}_{\mathrm{O}(3)} = \{\mathsf{0}\}$. In physical language, the piezoelectricity class $[\mathrm{O}(3)]$ describes materials that are not piezoelectric at all.

The first component consists of pseudo-vectors. Its basis tensors are

$$\mathsf{d}^p = \vec{e}^p \otimes g^0_{0[1,1]}, \qquad 1 \le p \le 3 .$$

In the compressed matrix notation, the nonzero entries of the matrices d^p become

$$e^p_{p1} = e^p_{p2} = e^p_{p3} = \frac{1}{\sqrt{3}} . \tag{3.2}$$

The second component also consists of pseudo-vectors. Its basis tensors are

$$\mathsf{d}^p = \sum_{i=-1}^{1} \sum_{j=-2}^{2} g^{p-5[i,j]}_{1[1,2]} \vec{e}^i \otimes g^j_{2[1,1]}, \qquad 4 \le p \le 6 .$$

Calculation of the Godunov–Gordienko matrices $g^{k[i,j]}_{1[1,2]}$, $-1 \le k \le 1$, gives the following results:

$$\begin{aligned}
\mathsf{d}^4 &= -\frac{1}{\sqrt{10}} \vec{e}^{-1} \otimes g^0_{2[1,1]} + \frac{\sqrt{3}}{\sqrt{10}} \vec{e}^{-1} \otimes g^2_{2[1,1]} + \frac{\sqrt{3}}{\sqrt{10}} \vec{e}^0 \otimes g^{-1}_{2[1,1]} - \frac{\sqrt{3}}{\sqrt{10}} \vec{e}^1 \otimes g^{-2}_{2[1,1]} , \\
\mathsf{d}^5 &= \frac{\sqrt{3}}{\sqrt{10}} \vec{e}^{-1} \otimes g^{-1}_{2[1,1]} + \frac{\sqrt{2}}{\sqrt{5}} \vec{e}^0 \otimes g^0_{2[1,1]} + \frac{\sqrt{3}}{\sqrt{10}} \vec{e}^1 \otimes g^1_{2[1,1]} , \\
\mathsf{d}^6 &= -\frac{\sqrt{3}}{\sqrt{10}} \vec{e}^{-1} \otimes g^{-2}_{2[1,1]} + \frac{\sqrt{3}}{\sqrt{10}} \vec{e}^0 \otimes g^1_{2[1,1]} - \frac{1}{\sqrt{10}} \vec{e}^1 \otimes g^0_{2[1,1]} - \frac{\sqrt{3}}{\sqrt{10}} \vec{e}^1 \otimes g^2_{2[1,1]} .
\end{aligned}$$

We calculate the nonzero elements of the matrices d^p, $4 \le p \le 6$, using (3.1). We obtain

$$\begin{aligned}
e^4_{11} &= \frac{2}{\sqrt{15}}, & e^4_{12} = e^4_{13} &= -\frac{1}{\sqrt{15}}, & e^4_{26} = e^4_{35} &= \frac{\sqrt{3}}{\sqrt{10}} , \\
e^5_{22} &= \frac{2}{\sqrt{15}}, & e^5_{16} = e^5_{34} &= \frac{\sqrt{3}}{\sqrt{10}}, & e^5_{21} = e^5_{23} &= -\frac{1}{\sqrt{15}} , \\
e^6_{33} &= \frac{2}{\sqrt{15}}, & e^6_{15} = e^6_{24} &= \frac{\sqrt{3}}{\sqrt{10}}, & e^6_{31} = e^6_{32} &= -\frac{1}{\sqrt{15}} .
\end{aligned} \tag{3.3}$$

The third component consists of pseudo-deviators. Its basis tensors are

$$\mathsf{d}^p = \sum_{i=-1}^{1} \sum_{j=-2}^{2} g^{p-9[i,j]}_{2[1,2]} \vec{e}^i \otimes g^j_{2[1,1]}, \qquad 7 \le p \le 11 .$$

Calculation of the Godunov–Gordienko matrices $g^{k[i,j]}_{2[1,2]}$, $-2 \leq k \leq 2$, gives the following results:

$$
\begin{aligned}
\mathsf{d}^7 &= \frac{1}{\sqrt{6}}\vec{e}^{-1} \otimes g^{-1}_{2[1,1]} - \frac{\sqrt{2}}{\sqrt{3}}\vec{e}^{0} \otimes g^{2}_{2[1,1]} - \frac{1}{\sqrt{6}}\vec{e}^{1} \otimes g^{1}_{2[1,1]}, \\
\mathsf{d}^8 &= -\frac{1}{\sqrt{6}}\vec{e}^{-1} \otimes g^{-2}_{2[1,1]} - \frac{1}{\sqrt{6}}\vec{e}^{0} \otimes g^{1}_{2[1,1]} + \frac{1}{\sqrt{2}}\vec{e}^{1} \otimes g^{0}_{2[1,1]} - \frac{1}{\sqrt{6}}\vec{e}^{1} \otimes g^{2}_{2[1,1]}, \\
\mathsf{d}^9 &= \frac{1}{\sqrt{2}}\vec{e}^{-1} \otimes g^{1}_{2[1,1]} - \frac{1}{\sqrt{2}}\vec{e}^{1} \otimes g^{-1}_{2[1,1]}, \\
\mathsf{d}^{10} &= -\frac{1}{\sqrt{2}}\vec{e}^{-1} \otimes g^{0}_{2[1,1]} - \frac{1}{\sqrt{6}}\vec{e}^{-1} \otimes g^{2}_{2[1,1]} + \frac{1}{\sqrt{6}}\vec{e}^{0} \otimes g^{-1}_{2[1,1]} + \frac{1}{\sqrt{6}}\vec{e}^{1} \otimes g^{-2}_{2[1,1]}, \\
\mathsf{d}^{11} &= \frac{1}{\sqrt{6}}\vec{e}^{-1} \otimes g^{1}_{2[1,1]} + \frac{\sqrt{2}}{\sqrt{3}}\vec{e}^{0} \otimes g^{-2}_{2[1,1]} + \frac{1}{\sqrt{6}}\vec{e}^{1} \otimes g^{-1}_{2[1,1]}.
\end{aligned}
$$

The nonzero elements of the matrices d^p, $7 \leq p \leq 11$, become

$$
\begin{aligned}
&e^7_{16} = \frac{1}{\sqrt{6}}, \quad e^7_{21} = -\frac{1}{\sqrt{3}}, \quad e^7_{23} = \frac{1}{\sqrt{3}}, \quad e^7_{34} = -\frac{1}{\sqrt{6}}, \\
&e^8_{15} = \frac{1}{\sqrt{6}}, \quad e^8_{24} = -\frac{1}{\sqrt{6}}, \quad e^8_{31} = -\frac{1}{\sqrt{3}}, \quad e^8_{32} = \frac{1}{\sqrt{3}}, \\
&e^9_{14} = \frac{1}{\sqrt{2}}, \quad e^9_{36} = -\frac{1}{\sqrt{2}}, \\
&e^{10}_{12} = -\frac{1}{\sqrt{3}}, \quad e^{10}_{13} = \frac{1}{\sqrt{3}}, \quad e^{10}_{26} = \frac{1}{\sqrt{6}}, \quad e^{10}_{35} = -\frac{1}{\sqrt{6}}, \\
&e^{11}_{14} = \frac{1}{\sqrt{6}}, \quad e^{10}_{25} = -\frac{\sqrt{2}}{\sqrt{3}}, \quad e^{10}_{36} = \frac{1}{\sqrt{6}}.
\end{aligned}
\tag{3.4}
$$

The fourth component consists of *3rd order pseudo-deviators*. Its basis tensors are

$$
\mathsf{d}^p = \sum_{i=-1}^{1} \sum_{j=-2}^{2} g^{p-15[i,j]}_{3[1,2]} \vec{e}^i \otimes g^j_{2[1,1]}, \qquad 12 \leq p \leq 18.
$$

Calculation of the Godunov–Gordienko matrices $g^{k[i,j]}_{3[1,2]}$, $-3 \leq k \leq 3$, gives the following results:

$$
\begin{aligned}
\mathsf{d}^{12} &= -\frac{1}{\sqrt{2}}\vec{e}^{-1}\otimes g^{2}_{2[1,1]} - \frac{1}{\sqrt{2}}\vec{e}^{1}\otimes g^{-2}_{2[1,1]},\\
\mathsf{d}^{13} &= -\frac{1}{\sqrt{3}}\vec{e}^{-1}\otimes g^{1}_{2[1,1]} + \frac{1}{\sqrt{3}}\vec{e}^{0}\otimes g^{-2}_{2[1,1]} - \frac{1}{\sqrt{3}}\vec{e}^{1}\otimes g^{-1}_{2[1,1]},\\
\mathsf{d}^{14} &= \frac{\sqrt{2}}{\sqrt{5}}\vec{e}^{-1}\otimes g^{0}_{2[1,1]} - \frac{1}{\sqrt{30}}\vec{e}^{-1}\otimes g^{2}_{2[1,1]} + \frac{2\sqrt{2}}{\sqrt{15}}\vec{e}^{0}\otimes g^{-1}_{2[1,1]} + \frac{1}{\sqrt{30}}\vec{e}^{1}\otimes g^{-2}_{2[1,1]},\\
\mathsf{d}^{15} &= -\frac{1}{\sqrt{5}}\vec{e}^{-1}\otimes g^{-1}_{2[1,1]} + \frac{\sqrt{3}}{\sqrt{5}}\vec{e}^{0}\otimes g^{0}_{2[1,1]} - \frac{1}{\sqrt{5}}\vec{e}^{1}\otimes g^{1}_{2[1,1]},\\
\mathsf{d}^{16} &= \frac{1}{\sqrt{30}}\vec{e}^{-1}\otimes g^{-2}_{2[1,1]} + \frac{2\sqrt{2}}{\sqrt{15}}\vec{e}^{0}\otimes g^{1}_{2[1,1]} + \frac{\sqrt{2}}{\sqrt{5}}\vec{e}^{1}\otimes g^{0}_{2[1,1]} + \frac{1}{\sqrt{30}}\vec{e}^{1}\otimes g^{2}_{2[1,1]},\\
\mathsf{d}^{17} &= \frac{1}{\sqrt{3}}\vec{e}^{-1}\otimes g^{-1}_{2[1,1]} + \frac{1}{\sqrt{3}}\vec{e}^{0}\otimes g^{2}_{2[1,1]} - \frac{1}{\sqrt{3}}\vec{e}^{1}\otimes g^{1}_{2[1,1]},\\
\mathsf{d}^{18} &= \frac{1}{\sqrt{2}}\vec{e}^{-1}\otimes g^{-2}_{2[1,1]} - \frac{1}{\sqrt{2}}\vec{e}^{1}\otimes g^{2}_{2[1,1]}.
\end{aligned}
\tag{3.5}
$$

The nonzero elements of the matrices d^p, $12 \le p \le 18$, become

$$
\begin{aligned}
&e^{12}_{11} = -\frac{1}{2}, \quad e^{12}_{13} = \frac{1}{2}, \quad e^{12}_{35} = \frac{1}{\sqrt{2}},\\
&e^{13}_{14} = -\frac{1}{\sqrt{3}}, \quad e^{13}_{25} = -\frac{1}{\sqrt{3}}, \quad e^{13}_{36} = -\frac{1}{\sqrt{3}},\\
&e^{14}_{11} = -\frac{\sqrt{3}}{2\sqrt{5}}, \quad e^{14}_{12} = \frac{2}{\sqrt{15}}, \quad e^{14}_{13} = -\frac{1}{2\sqrt{15}}, \quad e^{14}_{26} = \frac{2\sqrt{2}}{\sqrt{15}}, \quad e^{14}_{35} = -\frac{1}{\sqrt{30}},\\
&e^{15}_{16} = -\frac{1}{\sqrt{5}}, \quad e^{15}_{21} = -\frac{1}{\sqrt{10}}, \quad e^{15}_{22} = \frac{\sqrt{2}}{\sqrt{5}}, \quad e^{15}_{23} = -\frac{1}{\sqrt{10}}, \quad e^{15}_{34} = -\frac{1}{\sqrt{5}},\\
&e^{16}_{15} = -\frac{1}{\sqrt{30}}, \quad e^{16}_{24} = \frac{2\sqrt{2}}{\sqrt{15}}, \quad e^{16}_{31} = -\frac{1}{2\sqrt{15}}, \quad e^{16}_{32} = \frac{2}{\sqrt{15}}, \quad e^{16}_{33} = -\frac{\sqrt{3}}{2\sqrt{5}},\\
&e^{17}_{16} = \frac{1}{\sqrt{3}}, \quad e^{17}_{21} = \frac{1}{\sqrt{6}}, \quad e^{17}_{23} = -\frac{1}{\sqrt{6}}, \quad e^{17}_{34} = -\frac{1}{\sqrt{3}},\\
&e^{18}_{15} = -\frac{1}{\sqrt{2}}, \quad e^{18}_{31} = -\frac{1}{2}, \quad e^{18}_{33} = \frac{1}{2}.
\end{aligned}
$$

The description of the corresponding homogeneous and isotropic random fields is given in [20]. In the following section, we consider another example.

3.3 Example: The Piezoelectricity Class $[D_2]$

The first question we have to answer is the following: what is the dimension of the space V_{D_2}? Before giving the answer, we need to explain an important notion from representation theory.

In what follows, we extensively use the book [2], which works with *complex* representations. Let V be a complex irreducible representation of a compact group K. According to [1], there are three mutually exclusive cases.

1. There is a map $j\colon \mathsf{V} \to \mathsf{V}$ such that $g(j\mathsf{v}) = j(g\mathsf{v})$, $j(z\mathsf{v}) = \overline{z}(j\mathsf{v})$ ($z \in \mathbb{C}$), and $j^2\mathsf{v} = \mathsf{v}$. Such a representation is said to be of *real type*. In this case, we split V into the $+1$ and -1 eigenspaces of j; these are irreducible real representations which are isomorphic: the invertible intertwining operator is multiplication by i. We denote each of the obtained real representations by the same symbol V.
2. There is a map $j\colon \mathsf{V} \to \mathsf{V}$ which has all but the last property. The missed property is modified as follows: $j^2\mathsf{v} = -\mathsf{v}$. Such a representation is said to be of *quaternionic type* and will never occur in our studies.
3. No such map exists. These representations are said to be of *complex type*. Let V be an irreducible representation of complex type, and suppose $t\mathsf{V}$ has the same underlying set and the same group action as V, but $z \in \mathbb{C}$ acts on $t\mathsf{V}$ as $\overline{z}$ used to act on V. The representation $t\mathsf{V}$ is called *conjugate* to V, it is irreducible and not equivalent to V. By definition, the representation $r\mathsf{V}$ has the same underlying space as V and the same action of K, but is regarded as a real linear space. This representation is irreducible.

Define $cH_1 = \mathbb{C} \otimes_{\mathbb{R}} H_1$. This is a complex linear space where the scalar-vector multiplication is defined by $z(z' \otimes \mathsf{v}) = (zz') \otimes \mathsf{v}$. This is an irreducible complex representation of $\mathrm{O}(3)$ where the group acts by $g(z \otimes \mathsf{v}) = z \otimes (g\mathsf{v})$.

Consider the restriction of the representation cH_1 to the subgroup $K = D_2$. By [2, Table 22.10], it is equivalent to $B_1 \oplus B_2 \oplus B_3$. Unless otherwise stated, all representations are of real type. On which subspaces of $\mathbb{R}^3$ does each of the components act?

[2, Table 3.1] contains the Euler angles of all elements E, C_{2x}, C_{2y} and C_{2z} of the group D_2. Using the definition of the Euler angles, we see that $E = \begin{pmatrix} 1&0&0\\0&1&0\\0&0&1 \end{pmatrix}$, $C_{2x} = \begin{pmatrix} 1&0&0\\0&-1&0\\0&0&-1 \end{pmatrix}$, $C_{2y} = \begin{pmatrix} -1&0&0\\0&1&0\\0&0&-1 \end{pmatrix}$, $C_{2z} = \begin{pmatrix} -1&0&0\\0&-1&0\\0&0&1 \end{pmatrix}$. We see that the restriction of the representation $B_1 \oplus B_2 \oplus B_3$ to the x-axis takes value 1 on E and C_{2x} and -1 on C_{2y} and C_{2z}. By [2, Table 22.4], this representation is B_3. Similarly, B_2 acts on the y-axis, and B_1 on the z-axis.

The irreducible components of the symmetric tensor square $\mathsf{S}^2(B_1 \oplus B_2 \oplus B_3)$ are calculated using [2, Table 22.8]. We obtain: three copies of the trivial representation A act on the one-dimensional spaces generated by the matrices $\vec{e}^1 \otimes \vec{e}^1$, $\vec{e}^2 \otimes \vec{e}^2$ and $\vec{e}^3 \otimes \vec{e}^3$. The representation B_1 acts on the one-dimensional space generated by the matrix $\frac{1}{\sqrt{2}}(\vec{e}^1 \otimes \vec{e}^2 + \vec{e}^2 \otimes \vec{e}^1)$, the representation B_2 on the space generated by $\frac{1}{\sqrt{2}}(\vec{e}^1 \otimes \vec{e}^3 + \vec{e}^3 \otimes \vec{e}^1)$, and the representation B_3 on the space generated by $\frac{1}{\sqrt{2}}(\vec{e}^2 \otimes \vec{e}^3 + \vec{e}^3 \otimes \vec{e}^2)$. Again by [2, Table 22.8] we have: the three copies of the trivial representation act on the one-dimensional spaces generated by the matrices $\frac{1}{\sqrt{2}}\vec{e}^1 \otimes (\vec{e}^2 \otimes \vec{e}^3 + \vec{e}^3 \otimes \vec{e}^2)$, $\frac{1}{\sqrt{2}}\vec{e}^2 \otimes (\vec{e}^1 \otimes \vec{e}^3 + \vec{e}^3 \otimes \vec{e}^1)$ and $\frac{1}{\sqrt{2}}\vec{e}^3 \otimes (\vec{e}^1 \otimes \vec{e}^2 + \vec{e}^2 \otimes \vec{e}^1)$. In the compressed matrix notation, they become 3×6 matrices with the following nonzero elements:

$$\mathsf{d}^1_{16} = \mathsf{d}^2_{25} = \mathsf{d}^3_{34} = 1\,. \tag{3.6}$$

In particular, $\dim \mathsf{V}_{D_2} = 3$ and $\theta = 3A$.

The normaliser of the group D_2 is $O \times Z^2_c$. Besides the value of $K = D_2$ considered above, the group K can also take values D_4, $D_2 \times Z^c_2$, D^h_4, $\mathcal{T}$, $D_4 \times Z^c_2$, O, $\mathcal{T} \times Z^c_2$, O^- and $O \times Z^c_2$. This follows from [2, Graph 11].

Put $K = D_2 \times Z^c_2$. By [24, Appendix B], the Schoenflies notation for K is D_{2h}. By [2, Table 31.5], the restriction of the representation cH_1 of the group O(3) to the subgroup $D_2 \times Z^c_2$ is equivalent to the direct sum $B_{1u} \oplus B_{2u} \oplus B_{3u}$, where B_{1u} acts on the z-axis, B_{2u} on the y-axis, and B_{3u} on the x-axis. The representation $\mathsf{S}^2(B_{1u} \oplus B_{2u} \oplus B_{3u})$ is equivalent to the direct sum $3A_g \oplus B_{1g} \oplus B_{2g} \oplus B_{3g}$, where the first copy of A_g acts on the one-dimensional space generated by the matrix $\vec{e}^1 \otimes \vec{e}^1$, the second copy on the space generated by $\vec{e}^2 \otimes \vec{e}^2$, and the third copy on the space generated by $\vec{e}^3 \otimes \vec{e}^3$. The representation B_{1g} acts on the space generated by the matrix $\frac{1}{\sqrt{2}}(\vec{e}^1 \otimes \vec{e}^1 + \vec{e}^2 \otimes \vec{e}^2)$, the representation B_{2g} on the space generated by $\frac{1}{\sqrt{2}}(\vec{e}^1 \otimes \vec{e}^3 + \vec{e}^3 \otimes \vec{e}^1)$, and the representation B_{3g} on the space generated by $\frac{1}{\sqrt{2}}(\vec{e}^2 \otimes \vec{e}^3 + \vec{e}^3 \otimes \vec{e}^2)$. In the space P, the basis tensors of the subspace V_{D_2} are the tensors (3.6), where the one-dimensional space generated by d^1 carries the representation $B_{3u} \otimes B_{3g} = A_u$, the space generated by d^2 carries the representation $B_{2u} \otimes B_{2g} = A_u$, and the space generated by d^3 carries the representation $B_{1u} \otimes B_{1g} = A_u$. In particular, $\theta = 3A_u$.

Put $K = D_4$. We proceed in the same manner. By [2, Table 24.10], the restriction of the representation cH^1 of the group O(3) to the subgroup K is equivalent to $A_2 \oplus E$. [2, Table 24.5] shows that $\mathsf{S}^2(A_2 \oplus E) = 2A_1 \oplus B_1 \oplus B_2 \oplus E$, where the first copy of the trivial representation A_1 acts on the one-dimensional space generated by the matrix $\frac{1}{\sqrt{2}}(\vec{e}^1 \otimes \vec{e}^1 + \vec{e}^2 \otimes \vec{e}^2)$, the second copy on the space generated by $\vec{e}^3 \otimes \vec{e}^3$. The representation B_1 acts on the space generated by the matrix $\frac{1}{\sqrt{2}}(\vec{e}^1 \otimes \vec{e}^1 - \vec{e}^2 \otimes \vec{e}^2)$, the representation B_2 on the space generated by $\frac{1}{\sqrt{2}}(\vec{e}^1 \otimes \vec{e}^2 + \vec{e}^2 \otimes \vec{e}^1)$, and the representation E on the two-dimensional space generated by $\frac{1}{\sqrt{2}}(\vec{e}^1 \otimes \vec{e}^3 + \vec{e}^3 \otimes \vec{e}^1)$ and $\frac{1}{\sqrt{2}}(\vec{e}^2 \otimes \vec{e}^3 + \vec{e}^3 \otimes \vec{e}^2)$. We see also that the irreducible component A_2 of the representation $A_2 \oplus E$ acts on the z-axis, while E acts on the (x, y)-plane.

We investigate the structure of the representation $(A_2 \oplus E) \otimes (2A_1 \oplus B_1 \oplus B_2 \oplus E)$, using [2, Table 24.8]. Using the above described technique, it is possible to write down the basis tensors in each irreducible component of the above 18-dimensional representation and find components acting on linear subspaces of the space V_{D_2}. This is easy, because the basis tensors (3.6) of this space have already been calculated.

It turns out that the above components are the components A_1 and B_1 of the tensor square $E \otimes E$ and the tensor product $A_2 \otimes B_2 = B_1$. The last equality follows from [2, Table 24.8].

Indeed, the component A_1 acts on the space generated by the tensor $\frac{1}{2}[\vec{e}^1 \otimes (\vec{e}^2 \otimes \vec{e}^3 + \vec{e}^3 \otimes \vec{e}^2) + \vec{e}^2 \otimes (\vec{e}^1 \otimes \vec{e}^3 + \vec{e}^3 \otimes \vec{e}^1)]$, the first B_1 component on the space generated by $\frac{1}{2}[\vec{e}^1 \otimes (\vec{e}^2 \otimes \vec{e}^3 + \vec{e}^3 \otimes \vec{e}^2) - \vec{e}^2 \otimes (\vec{e}^1 \otimes \vec{e}^3 + \vec{e}^3 \otimes \vec{e}^1)]$, and the second B_1 component on the space generated by $\frac{1}{\sqrt{2}}\vec{e}^3 \otimes (\vec{e}^1 \otimes \vec{e}^2 + \vec{e}^2 \otimes \vec{e}^1)$. In the compressed matrix notation, they become 3×6 matrices with the following nonzero elements:

$$\mathsf{d}^1_{16} = \mathsf{d}^1_{25} = \mathsf{d}^2_{16} = \frac{1}{\sqrt{2}}, \qquad \mathsf{d}^2_{25} = -\frac{1}{\sqrt{2}}, \qquad \mathsf{d}^3_{34} = 1\,. \tag{3.7}$$

Moreover, $\theta = A_1 \oplus 2B_1$.

Put $K = D^h_4$. By [24, Appendix B], the Schoenflies notation for K is D_{2d}. By [2, Table 42.5], the restriction of the representation cH_1 of the group $\mathrm{O}(3)$ to the subgroup D^h_4 is equivalent to the direct sum $B_2 \oplus E$, where B_2 acts on the z-axis, and E acts on the (x, y)-plane. The symmetric tensor square $\mathsf{S}^2(B_2 \oplus E)$ is equivalent to the direct sum $2A_1 \oplus B_1 \oplus B_2 \oplus E$, where the first copy of A_1 acts on the one-dimensional space generated by the matrix $\frac{1}{\sqrt{2}}(\vec{e}^1 \otimes \vec{e}^1 + \vec{e}^2 \otimes \vec{e}^2)$, the second copy on the space generated by $\vec{e}^3 \otimes \vec{e}^3$, the representation B_1 on the space generated by $\frac{1}{\sqrt{2}}(\vec{e}^1 \otimes \vec{e}^1 - \vec{e}^2 \otimes \vec{e}^2)$, the representation B_2 on the space generated by $\frac{1}{\sqrt{2}}(\vec{e}^1 \otimes \vec{e}^2 + \vec{e}^2 \otimes \vec{e}^1)$, and the representation E on the two-dimensional space generated by $\frac{1}{\sqrt{2}}(\vec{e}^1 \otimes \vec{e}^3 + \vec{e}^3 \otimes \vec{e}^1)$ and $\frac{1}{\sqrt{2}}(\vec{e}^2 \otimes \vec{e}^3 + \vec{e}^3 \otimes \vec{e}^2)$. The irreducible components of the tensor product $(B_2 \oplus E) \otimes (2A_1 \oplus B_1 \oplus B_2 \oplus E)$ which act on linear subspaces of the space V_{D_2} are the component A_1 of the tensor product $E \otimes E$ that acts on the space generated by the tensor d^1 of Eq. (3.7), the component B_1 of the same tensor product that acts on the space generated by the tensor d^2 of Eq. (3.7), and the component $B_2 \otimes B_2 = A_1$ that acts on the space generated by the tensor d^3 of the above equation. We have $\theta = 2A_1 \oplus B_1$.

Put $K = D_4 \times Z^c_2$. By [24, Appendix B], the Schoenflies notation for K is D_{4h}. By [2, Table 33.5], the restriction of the representation cH_1 of the group $\mathrm{O}(3)$ to the subgroup $D_4 \times Z^c_2$ is equivalent to the direct sum $A_{2u} \oplus E_u$, where A_{2u} acts on the z-axis, and E_u acts on the (x, y)-plane. The symmetric tensor square $\mathsf{S}^2(A_{2u} \oplus E_u)$ is equivalent to the direct sum $2A_{1g} \oplus B_{1g} \oplus B_{2g} \oplus E_g$, where the first copy of A_{1g} acts on the one-dimensional space generated by the matrix $\frac{1}{\sqrt{2}}(\vec{e}^1 \otimes \vec{e}^1 + \vec{e}^2 \otimes \vec{e}^2)$, the second copy on the space generated by $\vec{e}^3 \otimes \vec{e}^3$, the representation B_{1g} on the space generated by $\frac{1}{\sqrt{2}}(\vec{e}^1 \otimes \vec{e}^1 - \vec{e}^2 \otimes \vec{e}^2)$, the representation B_{2g} on the space generated by $\frac{1}{\sqrt{2}}(\vec{e}^1 \otimes \vec{e}^2 + \vec{e}^2 \otimes \vec{e}^1)$, and the representation E_g on the two-dimensional space generated by $\frac{1}{\sqrt{2}}(\vec{e}^1 \otimes \vec{e}^3 + \vec{e}^3 \otimes \vec{e}^1)$ and $\frac{1}{\sqrt{2}}(\vec{e}^2 \otimes \vec{e}^3 + \vec{e}^3 \otimes \vec{e}^2)$. The irreducible components of the tensor product $(A_{2u} \oplus E_u) \otimes (2A_{1g} \oplus B_{1g} \oplus B_{2g} \oplus E_g)$ which act on linear subspaces of the space V_{D_2} are the component A_{1u}

of the tensor product $E_g \otimes E_u$ that acts on the space generated by the tensor d^1 of Eq. (3.7), the component B_{1u} of the same tensor product that acts on the space generated by the tensor d^2 of Eq. (3.7), and the component $A_{2u} \otimes B_{1g} = B_{2u}$ that acts on the space generated by the tensor d^3 of the above equation. We have $\theta = A_{1u} \oplus B_{1u} \oplus B_{2u}$.

Put $K = \mathcal{T}$. By [2, Table 70.5], the restriction of the representation cH_1 of the group O(3) to the subgroup $\mathcal{T}$ is T. By [2, Table 70.5], the symmetric tensor product $\mathsf{S}^2(T)$ is equivalent to the direct sum $A \oplus {}^1E \oplus {}^2E \oplus T$. The representations ${}^1E$ and ${}^2E$ are of complex type and mutually conjugate. Denote $r\,{}^1E$ by E. We obtain: $\mathsf{S}^2(T)$ is equivalent to the direct sum $A \oplus E \oplus T$. The representation A acts on the one-dimensional space generated by the matrix $\frac{1}{\sqrt{3}}(\vec{e}^1 \otimes \vec{e}^1 + \vec{e}^2 \otimes \vec{e}^2 + \vec{e}^3 \otimes \vec{e}^3)$. The representation E acts on the two-dimensional space generated by the matrices $\frac{1}{\sqrt{2}}(\vec{e}^1 \otimes \vec{e}^1 - \vec{e}^2 \otimes \vec{e}^2)$ and $\frac{1}{\sqrt{6}}(-\vec{e}^1 \otimes \vec{e}^1 - \vec{e}^2 \otimes \vec{e}^2 + 2\vec{e}^3 \otimes \vec{e}^3)$. The representation T acts on the three-dimensional space generated by $\frac{1}{\sqrt{2}}(\vec{e}^2 \otimes \vec{e}^3 + \vec{e}^3 \otimes \vec{e}^2)$, $\frac{1}{\sqrt{2}}(\vec{e}^1 \otimes \vec{e}^3 + \vec{e}^3 \otimes \vec{e}^1)$ and $\frac{1}{\sqrt{2}}(\vec{e}^1 \otimes \vec{e}^2 + \vec{e}^2 \otimes \vec{e}^1)$. The irreducible components of the tensor product $T \otimes (A \oplus E \oplus T)$ which act on linear subspaces of the space V_{D_2} are the components A and E of the tensor product $E \otimes E$. Indeed, the component A acts on the one-dimensional space generated by the tensor $\frac{1}{\sqrt{6}}[\vec{e}^1 \otimes (\vec{e}^2 \otimes \vec{e}^3 + \vec{e}^3 \otimes \vec{e}^2) + \vec{e}^2 \otimes (\vec{e}^1 \otimes \vec{e}^3 + \vec{e}^3 \otimes \vec{e}^1) + \vec{e}^3 \otimes (\vec{e}^1 \otimes \vec{e}^2 + \vec{e}^2 \otimes \vec{e}^1)]$, while the component E acts on the two-dimensional space generated by the tensors $\frac{1}{2}[\vec{e}^1 \otimes (\vec{e}^2 \otimes \vec{e}^3 + \vec{e}^3 \otimes \vec{e}^2) - \vec{e}^2 \otimes (\vec{e}^1 \otimes \vec{e}^3 + \vec{e}^3 \otimes \vec{e}^1)]$ and $\frac{1}{2\sqrt{3}}[-\vec{e}^1 \otimes (\vec{e}^2 \otimes \vec{e}^3 + \vec{e}^3 \otimes \vec{e}^2) - \vec{e}^2 \otimes (\vec{e}^1 \otimes \vec{e}^3 + \vec{e}^3 \otimes \vec{e}^1) + 2\vec{e}^3 \otimes (\vec{e}^1 \otimes \vec{e}^2 + \vec{e}^2 \otimes \vec{e}^1)]$. In the compressed matrix notation, these tensors become the matrices with the following nonzero elements.

$$\mathsf{d}^1_{16} = \mathsf{d}^1_{25} = \mathsf{d}^1_{34} = \frac{1}{\sqrt{3}}, \quad \mathsf{d}^2_{16} = \frac{1}{\sqrt{2}}, \quad \mathsf{d}^2_{25} = -\frac{1}{\sqrt{2}}, \quad \mathsf{d}^3_{16} = \mathsf{d}^3_{25} = -\frac{1}{\sqrt{6}},$$
$$\mathsf{d}^3_{34} = \frac{\sqrt{2}}{\sqrt{3}}. \tag{3.8}$$

We have $\theta = A \oplus E$.

Put $K = \mathcal{T} \times Z^c_2$. By [24, Appendix B], the Schoenflies notation for K is T_h. By [2, Table 72.5], the restriction of the representation cH_1 of the group O(3) to the subgroup $\mathcal{T} \times Z^c_2$ is T_u. The symmetric tensor square $\mathsf{S}^2(T_u)$ is equivalent to the direct sum of complex irreducible representations $A_g \oplus {}^1E_g \oplus {}^2E_g \oplus T_g$ or of real irreducible representations $A_g \oplus E_g \oplus T_g$, where $E_g = r\,{}^1E_g$. The representation A_g acts on the one-dimensional space generated by the matrix $\frac{1}{\sqrt{3}}(\vec{e}^1 \otimes \vec{e}^1 + \vec{e}^2 \otimes \vec{e}^2 + \vec{e}^3 \otimes \vec{e}^3)$. The representation E_g acts on the two-dimensional space generated by the matrices $\frac{1}{\sqrt{2}}(\vec{e}^1 \otimes \vec{e}^1 - \vec{e}^2 \otimes \vec{e}^2)$ and $\frac{1}{\sqrt{6}}(-\vec{e}^1 \otimes \vec{e}^1 - \vec{e}^2 \otimes \vec{e}^2 + 2\vec{e}^3 \otimes \vec{e}^3)$. The representation T_g acts on the three-dimensional space generated by $\frac{1}{\sqrt{2}}(\vec{e}^2 \otimes \vec{e}^3 + \vec{e}^3 \otimes \vec{e}^2)$, $\frac{1}{\sqrt{2}}(\vec{e}^1 \otimes \vec{e}^3 + \vec{e}^3 \otimes \vec{e}^1)$ and $\frac{1}{\sqrt{2}}(\vec{e}^1 \otimes \vec{e}^2 + \vec{e}^2 \otimes \vec{e}^1)$. The irreducible components of the tensor product $T_u \otimes (A_g \oplus E \oplus T_g)$ which act on linear subspaces of the space V_{D_2} are the components A_u and E_u of the tensor product $T_u \otimes T_g$, where

E_u is the result of applying the second construction to any of the irreducible unitary representations ${}^1E_u$ or ${}^2E_u$. Indeed, the component A_u acts on the one-dimensional space generated by the tensor d^1 of (3.8), while the component E_u acts on the two-dimensional space generated by the tensors d^2 and d^3 of (3.8). We have $\theta = A_u \oplus E_u$.

Put $K = \mathcal{O}$. By [2, Table 69.5], the restriction of the representation cH_1 of the group O(3) to the subgroup $\mathcal{T}$ is T_1. The symmetric tensor square $\mathsf{S}^2(T_1)$ is equivalent to the direct sum $A_1 \oplus E \oplus T_2$. The representation A_1 acts on the one-dimensional space generated by the matrix $\frac{1}{\sqrt{3}}(\vec{e}^1 \otimes \vec{e}^1 + \vec{e}^2 \otimes \vec{e}^2 + \vec{e}^3 \otimes \vec{e}^3)$. The representation E acts on the two-dimensional space generated by the matrices $\frac{1}{\sqrt{2}}(\vec{e}^1 \otimes \vec{e}^1 - \vec{e}^2 \otimes \vec{e}^2)$ and $\frac{1}{\sqrt{6}}(-\vec{e}^1 \otimes \vec{e}^1 - \vec{e}^2 \otimes \vec{e}^2 + 2\vec{e}^3 \otimes \vec{e}^3)$. The representation T_2 acts on the three-dimensional space generated by $\frac{1}{\sqrt{2}}(\vec{e}^2 \otimes \vec{e}^3 + \vec{e}^3 \otimes \vec{e}^2)$, $\frac{1}{\sqrt{2}}(\vec{e}^1 \otimes \vec{e}^3 + \vec{e}^3 \otimes \vec{e}^1)$ and $\frac{1}{\sqrt{2}}(\vec{e}^1 \otimes \vec{e}^2 + \vec{e}^2 \otimes \vec{e}^1)$. The irreducible components of the tensor product $T_1 \otimes (A_1 \oplus E \oplus T_2)$ which act on linear subspaces of the space V_{D_2} are the components A_2 and E of the tensor product $T_1 \otimes T_2$. Indeed, the component A_2 acts on the one-dimensional space generated by the tensor d^1 of (3.8), while the component E acts on the two-dimensional space generated by the tensors d^2 and d^3 of (3.8). We have $\theta = A_2 \oplus E$.

Put $K = \mathcal{O}^-$. By [24, Appendix B], the Schoenflies notation for K is T_d. By [2, Table 73.5], the restriction of the representation cH_1 of the group O(3) to the subgroup $\mathcal{O}^-$ is T_2. The symmetric tensor square $\mathsf{S}^2(T_2)$ is equivalent to the direct sum $A_1 \oplus E \oplus T_2$. The representation A_1 acts on the one-dimensional space generated by the matrix $\frac{1}{\sqrt{3}}(\vec{e}^1 \otimes \vec{e}^1 + \vec{e}^2 \otimes \vec{e}^2 + \vec{e}^3 \otimes \vec{e}^3)$. The representation E acts on the two-dimensional space generated by the matrices $\frac{1}{\sqrt{2}}(\vec{e}^1 \otimes \vec{e}^1 - \vec{e}^2 \otimes \vec{e}^2)$ and $\frac{1}{\sqrt{6}}(-\vec{e}^1 \otimes \vec{e}^1 - \vec{e}^2 \otimes \vec{e}^2 + 2\vec{e}^3 \otimes \vec{e}^3)$. The representation T_2 acts on the three-dimensional space generated by $\frac{1}{\sqrt{2}}(\vec{e}^2 \otimes \vec{e}^3 + \vec{e}^3 \otimes \vec{e}^2)$, $\frac{1}{\sqrt{2}}(\vec{e}^1 \otimes \vec{e}^3 + \vec{e}^3 \otimes \vec{e}^1)$ and $\frac{1}{\sqrt{2}}(\vec{e}^1 \otimes \vec{e}^2 + \vec{e}^2 \otimes \vec{e}^1)$. The irreducible components of the tensor product $T_2 \otimes (A_1 \oplus E \oplus T_2)$ which act on linear subspaces of the space V_{D_2} are the components A_1 and E of the tensor product $T_2 \otimes T_2$. Indeed, the component A_1 acts on the one-dimensional space generated by the tensor d^1 of (3.8), while the component E acts on the two-dimensional space generated by the tensors d^2 and d^3 of (3.8). We have $\theta = A_1 \oplus E$.

Finally, put $K = \mathcal{O} \times Z_2^c$. By [24, Appendix B], the Schoenflies notation for K is O_h. By [2, Table 73.5], the restriction of the representation rH_1 of the group O(3) to the subgroup $\mathcal{O} \times Z_2^c$ is T_{1u}. The symmetric tensor square $\mathsf{S}^2(T_{1u})$ is equivalent to the direct sum $A_{1g} \oplus E_g \oplus T_{2g}$. The representation A_{1g} acts on the one-dimensional space generated by the matrix $\frac{1}{\sqrt{3}}(\vec{e}^1 \otimes \vec{e}^1 + \vec{e}^2 \otimes \vec{e}^2 + \vec{e}^3 \otimes \vec{e}^3)$. The representation E_g acts on the two-dimensional space generated by the matrices $\frac{1}{\sqrt{2}}(\vec{e}^1 \otimes \vec{e}^1 - \vec{e}^2 \otimes \vec{e}^2)$ and $\frac{1}{\sqrt{6}}(-\vec{e}^1 \otimes \vec{e}^1 - \vec{e}^2 \otimes \vec{e}^2 + 2\vec{e}^3 \otimes \vec{e}^3)$. The representation T_{2g} acts on the three-dimensional space generated by $\frac{1}{\sqrt{2}}(\vec{e}^2 \otimes \vec{e}^3 + \vec{e}^3 \otimes \vec{e}^2)$, $\frac{1}{\sqrt{2}}(\vec{e}^1 \otimes \vec{e}^3 + \vec{e}^3 \otimes \vec{e}^1)$ and $\frac{1}{\sqrt{2}}(\vec{e}^1 \otimes \vec{e}^2 + \vec{e}^2 \otimes \vec{e}^1)$. The irreducible components of the tensor product $T_{1u} \otimes (A_{1g} \oplus E_g \oplus T_{2g})$ which act on linear subspaces of the space V_{D_2} are the components A_{2u} and E_u of the tensor product $T_{1u} \otimes T_{2g}$. Indeed,

Table 3.1 The bases of the space V_{D_2}

K	θ	The basis
D_2	$3A$	(3.6)
$D_2 \times Z_2^c$	$3A_u$	(3.6)
D_4	$A_1 \oplus 2B_1$	(3.7)
D_4^h	$2A_1 \oplus B_1$	(3.7)
$D_4 \times Z_2^c$	$A_{1u} \oplus B_{1u} \oplus B_{2u}$	(3.7)
$\mathcal{T}$	$A \oplus E$	(3.8)
$\mathcal{T} \times Z_2^c$	$A_u \oplus E_u$	(3.8)
$\mathcal{O}$	$A_2 \oplus E$	(3.8)
$\mathcal{O}^-$	$A_1 \oplus E$	(3.8)
$\mathcal{O} \times Z_2^c$	$A_{2u} \oplus E_u$	(3.8)

the component A_{2u} acts on the one-dimensional space generated by the tensor d^1 of (3.8), while the component E_u acts on the two-dimensional space generated by the tensors d^2 and d^3 of (3.8). We have $\theta = A_{2u} \oplus E_u$.

Our investigation is summed up in Table 3.1.

In a similar way, it is possible to find the bases in the remaining spaces V_G. For the reader's convenience, we list the normalisers of the piezoelectricity classes not considered here. The normaliser of the groups Z_2, Z_3, Z_2^-, Z_4^-, SO(2), O(2), O(2)$^-$ is O(2) $\times Z_2^c$. The normaliser of D_2^v and D_4^h is $D_4 \times Z_2^c$. The normaliser of D_3 and D_3^v is $D_6 \times Z_2^c$. The normaliser of D_h^6 is $D_6 \times Z_2^c$. Finally, the normaliser of $\mathcal{O}^-$ is $\mathcal{O} \times Z_2^c$.

Note that the groups D_2 and $D_2 \times Z_2^c$ belong to the *orthotropic crystal system*, the groups D_4, D_4^h and $D_4 \times Z_2^c$ to the *tetragonal crystal system*, and the rest of the groups to the *cubic crystal system*. It turns out that two groups share the same basis if and only if they belong to the same crystal system.

Chapter 4
Correlation Structures

4.1 Introduction

In this chapter, we calculate the one- and two-point correlation tensors of a homogeneous and (K, θ)-isotropic random field for all 10 cases in Table 3.1. The spectral expansion of the field will also be calculated.

The main idea is simple. We consider the correlation structure of a homogeneous random field and find the conditions under which it is isotropic.

4.2 Homogeneous Random Fields

There exists no complete description of the correlation structure of a homogeneous random field taking values in a *real* finite-dimensional linear space. Instead, we have the following partial result.

Let $\mathsf{d}^1, \mathsf{d}^2, \dots, \mathsf{d}^{\dim \mathsf{V}_G}$ be the basis of the space V_G constructed in Chap. 3.

Theorem 4.1 ([7]) *A* $c\mathsf{V}_G$*-valued random field* $\{\,\mathsf{d}(\vec{x})\colon \vec{x} \in \mathbb{R}^3\,\}$ *is homogeneous if and only if its two-point correlation tensor has the form*

$$\langle \mathsf{d}(\vec{x}), \mathsf{d}(\vec{y}) \rangle_{pq} = \int_{\hat{\mathbb{R}}^3} \mathrm{e}^{\mathrm{i}(\vec{k}, \vec{y}-\vec{x})}\, \mathrm{d}F_{pq}(\vec{k})\,. \tag{4.1}$$

If the above random field takes values in V_G, *then*

$$F_{pq}(-A) = F_{qp}(A), \qquad A \in \mathfrak{B}(\hat{\mathbb{R}}^3)\,, \tag{4.2}$$

where $-A = \{\,-\vec{k}\colon \vec{k} \in A\,\}$.

A. Malyarenko et al., *Random Fields of Piezoelectricity and Piezomagnetism*, SpringerBriefs in Mathematical Methods,
https://doi.org/10.1007/978-3-030-60064-8_4

In Theorem 4.1, $\hat{\mathbb{R}}^3$ is the *wavenumber domain*, and F_{pq} is a measure on the Borel σ-field $\mathfrak{B}(\hat{\mathbb{R}}^3)$ taking values in the set of Hermitian nonnegative-definite matrices with $\dim \mathsf{V}_G$ rows.

Thus, if one applies condition (4.2), no V_G-valued random fields are missing. Some of the remaining fields may still be $c\mathsf{V}_G$-valued and each case must be analysed separately.

4.3 Conditions for Isotropy

Let d^0 be the constant one-point correlation tensor of a homogeneous random field $\mathsf{d}(\vec{x})$: $\langle \mathsf{d}(\vec{x}) \rangle = \mathsf{d}^0$. Substitute this value into the first equation in (2.2). We obtain: a tensor $\mathsf{d}^0 \in \mathsf{P}$ is a one-point correlation tensor of a homogeneous and isotropic random field if and only if $\mathsf{d}^0 = \theta(g)\mathsf{d}^0$ for all $g \in K$. In other words, if the representation θ does not contain any copies of the trivial representation θ^0 of the group K, then $\langle \mathsf{d}(\vec{x}) \rangle = \mathsf{0}$. Otherwise, $\langle \mathsf{d}(\vec{x}) \rangle$ is an arbitrary tensor lying in the subspace V^0_G.

Example 4.1 Consider the random fields for which we constructed a basis in the space V_G in Chap. 3. When $G = Z_1$ and $K = \mathrm{O}(3)$, we have $\theta = 2\theta^1 \oplus \theta^2 \oplus \theta^3$. No trivial representations are here. It follows that $\langle \mathsf{d}(\vec{x}) \rangle = \mathsf{0}$. The same is true when $G = D_2$ and $K \in \{D_2 \times Z^c_2, D_4 \times Z^c_2, \mathcal{T} \times Z^c_2, \mathcal{O}, \mathcal{O} \times Z^c_2\}$.

When $K = D_2$, we have $\theta = 3A$ and

$$\langle \mathsf{d}(\vec{x}) \rangle = C_1\mathsf{d}^1 + C_2\mathsf{d}^2 + C_3\mathsf{d}^3, \qquad C_i \in \mathbb{R},$$

where d^i are the tensors of the basis (3.6).

When $K \in \{D_4, \mathcal{T}, \mathcal{O}^-\}$, the representation θ contains one copy of the trivial representation of K, and we have

$$\langle \mathsf{d}(\vec{x}) \rangle = C_1\mathsf{d}^1, \qquad C_1 \in \mathbb{R},$$

where d^1 is the tensor of Eq. (3.7) when $K = D_4$, and the tensor of Eq. (3.8) otherwise.

Finally, when $K = D^h_4$, the representation θ contains two copies of the trivial representation of K, and we have

$$\langle \mathsf{d}(\vec{x}) \rangle = C_1\mathsf{d}^1 + C_2\mathsf{d}^2, \qquad C_i \in \mathbb{R},$$

where d^i are the tensors of Eq. (3.7).

To investigate the two-point correlation tensor, we substitute (4.1) into the second equation in (2.2). After simple algebraic manipulations, we obtain the following result: Eq. (4.1) describes the two-point correlation tensor of a $c\mathsf{V}_G$-valued homogeneous and isotropic random field if and only if

$$F(gA) = (\theta \otimes \theta)(g)F(A), \qquad g \in K, \quad A \in \mathfrak{B}(\hat{\mathbb{R}}^3), \tag{4.3}$$

for all $g \in K$. For details, see [20].

We have two conditions for the measure $F = F_{pq}$: (4.2) and (4.3). Is it possible to replace them with *one* condition? The following lemma gives an affirmative answer to this question. To formulate the lemma, we need more notation.

Consider the representation $c\mathsf{V}_G \otimes \mathsf{V}_G$. By [1], it is a complex representation of real type. The $+1$-eigenspace of j is the real linear space of Hermitian operators in $c\mathsf{V}_G$ and the restriction of the representation $c\mathsf{V}_G \otimes \mathsf{V}_G$ to this space is a copy of $\mathsf{V}_G \otimes \mathsf{V}_G$.

Lemma 4.1 *There exists a group $\tilde{K}$ and its real orthogonal representation $\tilde{\mathsf{V}} \subseteq \mathsf{V}_G \otimes \mathsf{V}_G$ such that conditions (4.2) and (4.3) are equivalent to the following conditions.*

- *The measure F takes values in the intersection of the set of all nonnegative-definite Hermitian linear operators in $\mathsf{V}_G \otimes \mathsf{V}_G$ with the space $\tilde{\mathsf{V}}$.*
- $$F(\tilde{g}A) = \tilde{g}F(A), \qquad \tilde{g} \in \tilde{K}, \quad A \in \mathfrak{B}(\hat{\mathbb{R}}^3). \tag{4.4}$$

Proof This lemma was proved in [20]. Here we give a simplified proof.

Let K be a closed subgroup of the group $\mathrm{O}(d)$, and let $Z = \{E, -E\}$, where E is the $d \times d$ identity matrix. Let $\mathsf{V}_G \otimes \mathsf{V}_G^Z$ be the same set as $\mathsf{V}_G \otimes \mathsf{V}_G$, but suppose the group Z acts trivially on the subspace $\mathsf{S}^2(\mathsf{V}_G)$ of symmetric linear operators: $gA = A$ for all $g \in Z$, and nontrivially on the subspace $\Lambda^2(\mathsf{V}_G)$ of skew-symmetric linear operators: $-EA = -A$. We observe that condition (4.2) takes the form

$$F(gA) = gF(A), \qquad g \in Z, \quad A \in \mathfrak{B}(\hat{\mathbb{R}}^3). \tag{4.5}$$

Assume $-E \in K$. Let $\mathsf{V}_G \otimes \mathsf{V}_G^+$ (resp. $\mathsf{V}_G \otimes \mathsf{V}_G^-$) be the subspace of the space $\mathsf{V}_G \otimes \mathsf{V}_G$ where $-E$ acts trivially (resp. nontrivially). Put $\tilde{K} = K$ and

$$\tilde{\mathsf{V}} = (\mathsf{V}_G \otimes \mathsf{V}_G^+ \cap \mathsf{S}^2(\mathsf{V}_G)) \oplus (\mathsf{V}_G \otimes \mathsf{V}_G^- \cap \Lambda^2(\mathsf{V}_G)).$$

Indeed, this is the maximal subspace where the actions (4.3) and (4.5) coincide.

Assume $-E \notin K$. Put $\tilde{K} = K \cup \{-Eg : g \in K\}$. It is easy to check that $\tilde{K}$ is a closed subgroup of $\mathrm{O}(d)$. Let $\tilde{\mathsf{V}}$ be the same set as $\mathsf{V}_G \otimes \mathsf{V}_G$, but define the action of the group $\tilde{K}$ by

$$\tilde{g}A = \begin{cases} gA, & \text{if } \tilde{g} = g \in K, \\ -E\tilde{g}A, & \text{if } \tilde{g} \notin K, A \in \mathsf{S}^2(\mathsf{V}_G), \\ -gA, & \text{if } \tilde{g} \notin K, A \in \Lambda^2(\mathsf{V}_G). \end{cases}$$

It is easy to check that this formula defines an orthogonal representation of the group $\tilde{K}$ whose restriction to K is equal to $\mathsf{V}_G \otimes \mathsf{V}_G$, while the restriction to Z is equal to $\mathsf{V}_G \otimes \mathsf{V}_G^Z$.

Example 4.2 The groups O(3), $D_2 \times Z_2^c$, $D_4 \times Z_2^c$, $\mathcal{T} \times Z_2^c$, and $\mathcal{O} \times Z_2^c$ contain $-E$. Moreover, $\theta(-E) = E$, and $\tilde{\mathsf{V}_G} = \mathsf{S}^2(\mathsf{V}_G)$.

When $K = \mathrm{O}(3)$, we find

$$\tilde{\mathsf{V}} = 5H_0 \oplus 8H_1 \oplus 10H_2 \oplus 7H_3 \oplus 5H_4 \oplus 2H_5 \oplus H_6\,. \tag{4.6}$$

When $K = D_2 \times Z_2^c$, [2, Table 31.8] gives

$$\tilde{\theta} = \mathsf{S}^2(\theta) = \mathsf{S}^2(3A_u) = 6A_g\,. \tag{4.7}$$

When $K = D_4 \times Z_2^c$, [2, Table 33.8] gives

$$\tilde{\theta} = \mathsf{S}^2(\theta) = \mathsf{S}^2(A_{1u} \oplus B_{1u} \oplus B_{2u}) = 3A_{1g} \oplus A_{2g} \oplus B_{1g} \oplus B_{2g}\,. \tag{4.8}$$

When $K = \mathcal{T} \times Z_2^c$, [2, Table 72.8] gives

$$\tilde{\theta} = \mathsf{S}^2(\theta) = \mathsf{S}^2(A_u \oplus E_u) = 2A_g \oplus 2E_g\,. \tag{4.9}$$

When $K = \mathcal{O} \times Z_2^c$, [2, Table 71.8] gives

$$\tilde{\theta} = \mathsf{S}^2(\theta) = \mathsf{S}^2(A_{2u} \oplus E_u) = 2A_{1g} \oplus 2E_g\,. \tag{4.10}$$

The groups D_2, D_4, $\mathcal{T}$, and $\mathcal{O}$ are subgroups of the group SO(3). We have $\tilde{K} = K \times Z_c^2$, where $Z_2^c = \{E, -E\}$. Moreover,

$$\tilde{\theta} = \mathsf{S}^2(\theta) \otimes A_g \oplus \Lambda^2(\theta) \otimes A_u\,,$$

where A_g (resp. A_u) is the trivial (resp. nontrivial) irreducible representation of Z_2^c.

When $G = D_2$, we obtain $\mathsf{S}^2(\theta) = 6A$, $\Lambda^2(\theta) = 3A$, and

$$\tilde{\theta} = 6A \otimes A_g \oplus 3A \otimes A_u = 6A_g \oplus 3A_u\,. \tag{4.11}$$

When $G = D_4$, [2, Table 24.8] gives $\mathsf{S}^2(\theta) = 4A_1 \oplus 2B_1$, $\Lambda^2(\theta) = A_1 \oplus 2B_1$, and

$$\tilde{\theta} = (4A_1 \oplus 2B_1) \otimes A_g \oplus (A_1 \oplus 2B_1) \otimes A_u = 4A_{1g} \oplus 2B_{1g} \oplus A_{1u} \oplus 2B_{1u}\,. \tag{4.12}$$

When $G = \mathcal{T}$, [2, Table 70.8] gives $\mathsf{S}^2(\theta) = 2A \oplus 2E$, $\Lambda^2(\theta) = A \oplus E$, and

$$\tilde{\theta} = (2A \oplus 2E) \otimes A_g \oplus (A \oplus E) \otimes A_u = 2A_g \oplus 2E_g \oplus A_u \oplus E_u\,. \tag{4.13}$$

When $G = O$, [2, Table 71.8] gives $\mathsf{S}^2(\theta) = 2A_1 \oplus 2E$, $\Lambda^2(\theta) = A_2 \oplus E$, and

$$\tilde{\theta} = (2A_1 \oplus 2E) \otimes A_g \oplus (A_2 \oplus E) \otimes A_u = 2A_{1g} \oplus 2E_g \oplus A_{2u} \oplus E_u . \quad (4.14)$$

Finally, the groups D_4^h and O^- are not subgroups of the group SO(3) and do not contain $-E$. Put $K = D_4^h$. We have $\tilde{K} = D_4 \times Z_2^c$. By [2, Table 41.8], we have $\mathsf{S}^2(\theta) = 4A_1 \oplus 2B_1$, $\Lambda^2(\theta) = A_1 \oplus 2B_1$. By [2, Table 41.4], under the isomorphism $\pi \colon D_4^h \to D_4$ given by

$$\pi(g) = \begin{cases} g, & \text{if } g \in D_4^h \cap D^4 , \\ -g, & \text{otherwise} \end{cases}$$

the representation A_1 becomes A_2, and B_1 becomes B_2. Then

$$\tilde{\theta} = (4A_1 \oplus 2B_1) \otimes A_g \oplus (A_2 \oplus 2B_2) \otimes A_u = 4A_{1g} \oplus 2B_{1g} \oplus A_{2u} \oplus 2B_{2u} . \quad (4.15)$$

Put $K = O^-$. We have $\tilde{K} = O \times Z_2^c$. By [2, Table 69.8], we have $\mathsf{S}^2(\theta) = A_1 \oplus A_2 \oplus 2E$, $\Lambda^2(\theta) = A_2 \oplus E$. By [2, Table 73.4], under the isomorphism $\pi \colon O^- \to O$ given by

$$\pi(g) = \begin{cases} g, & \text{if } g \in O^- \cap O , \\ -g, & \text{otherwise} \end{cases}$$

the representation A_2 becomes A_1, and E does not change. Then

$$\tilde{\theta} = (2A_1 \oplus 2E) \otimes A_g \oplus (A_1 \oplus E) \otimes A_u = 2A_{1g} \oplus 2E_g \oplus A_{1u} \oplus E_u . \quad (4.16)$$

Consider the measure

$$\nu(A) = F_{pp}(A), \qquad A \in \mathfrak{B}(\hat{\mathbb{R}}^3) .$$

It is well known that the measure F is absolutely continuous with respect to ν, that is, there exists a function f defined on $\hat{\mathbb{R}}^3$ taking values in the set of nonnegative-definite Hermitian matrices on V_G with unit trace such that

$$F_{pq}(A) = \int_A f_{pq}(\vec{k}) \, \mathrm{d}\nu(\vec{k}) ,$$

see [5]. Equation (4.1) becomes

$$\langle \mathsf{d}(\vec{x}), \mathsf{d}(\vec{y}) \rangle_{pq} = \int_{\hat{\mathbb{R}}^3} \mathrm{e}^{\mathrm{i}(\vec{k}, \vec{y} - \vec{x})} f_{pq}(\vec{k}) \, \mathrm{d}\nu(\vec{k}) , \quad (4.17)$$

while Eq. (4.4) becomes

$$\begin{aligned} \nu(\tilde{g}A) &= \nu(A)\,, \\ f(\tilde{g}\vec{k}) &= \tilde{g}f(\vec{k}) \end{aligned} \tag{4.18}$$

for all $\tilde{g} \in \tilde{K}$, $A \in \mathfrak{B}(\hat{\mathbb{R}}^3)$, and $\vec{k} \in \hat{\mathbb{R}}^3$.

4.4 The Structure of the Orbit Space

In this section, we analyse the first condition in (4.18). The description of all measures satisfying this condition is well known, see [6].

Under the action of $\tilde{K}$ on $\hat{\mathbb{R}}^3$ by matrix-vector multiplication, the wavenumber domain is divided into orbits. Let $\hat{\mathbb{R}}^3/\tilde{K}$ be the set of all orbits, and let $\pi\colon \hat{\mathbb{R}}^3 \to \hat{\mathbb{R}}^3/\tilde{K}$ be the mapping that maps a point $\vec{k} \in \hat{\mathbb{R}}^3$ to its orbit. A set $A \subseteq \hat{\mathbb{R}}^3/\tilde{K}$ is open if and only if $\pi^{-1}(A)$ is open in $\hat{\mathbb{R}}^3$. In what follows, for each group $\tilde{K}$ considered above we construct a subset of $\hat{\mathbb{R}}^3$ such that it intersects with each orbit at exactly one point, and the mapping that maps a point to the corresponding orbit is a homeomorphism. Denote this subset by the same symbol $\hat{\mathbb{R}}^3/\tilde{K}$.

It turns out that the set $\hat{\mathbb{R}}^3/\tilde{K}$ is a *finite stratified space*. This means that the above set is a union of disjoint strata $\{\, (\hat{\mathbb{R}}^3/\tilde{K})_m \colon 0 \leq m \leq M-1 \,\}$. Each stratum is a manifold, and the boundary of any stratum is a union of some other strata of a smaller dimension. Denote by $\vec{\lambda}_m$ a chart of the manifold $(\hat{\mathbb{R}}^3/\tilde{K})_m$.

The orbit of a point $\vec{\lambda}_m \in (\hat{\mathbb{R}}^3/\tilde{K})_m$ is also a manifold. Denote it by $O(\vec{\lambda}_m)$, and let $\vec{\sigma}_m$ be a chart on this orbit. It is well known that there exists a unique probabilistic $\tilde{K}$-invariant measure on $O(\vec{\lambda}_m)$, denote it by $\mathrm{d}\vec{\sigma}_m$. A measure ν satisfies the first condition in (4.18) if and only if there exist finite measures μ_m on Borel σ-fields $\mathfrak{B}((\hat{\mathbb{R}}^3/\tilde{K})_m)$ such that the restriction of ν to $\pi^{-1}((\hat{\mathbb{R}}^3/\tilde{K})_m)$ is the product of the measures μ_m and $\mathrm{d}\vec{\sigma}_m$. Equation (4.17) becomes

$$\langle \mathsf{d}(\vec{x}), \mathsf{d}(\vec{y})\rangle_{pq} = \sum_{m=0}^{M-1} \int_{(\hat{\mathbb{R}}^3/\tilde{K})_m} \int_{O(\vec{\lambda}_m)} \mathrm{e}^{\mathrm{i}(\vec{k},\vec{y}-\vec{x})} f_{pq}(\vec{k})\,\mathrm{d}\vec{\sigma}_m\,\mathrm{d}\mu_m(\vec{\lambda}_m)\,. \tag{4.19}$$

Example 4.3 Put $\tilde{K} = \mathrm{O}(3)$. The stratified space $\hat{\mathbb{R}}^3/\tilde{K}$ is a union of two strata

$$(\hat{\mathbb{R}}^3/\mathrm{O}(3))_0 = \{\vec{0}\}, \qquad (\hat{\mathbb{R}}^3/\mathrm{O}(3))_1 = \{\, (0,0,k_3)\colon k_3 > 0 \,\}\,.$$

The elements of the group O(3) that fix a point $\vec{\lambda}_m \in (\hat{\mathbb{R}}^3/\tilde{K})_m$ form a group K_m, the *stationary subgroup* of the point $\vec{\lambda}_m$. In fact, K_m depends only on m but not on $\vec{\lambda}_m$. We have $K_0 = \mathrm{O}(3)$ and $K_1 = \mathrm{O}(2)$. Moreover, for any group $\tilde{K}$ we have $(\hat{\mathbb{R}}^3/\tilde{K})_0 = \{\vec{0}\}$ and $K_0 = \tilde{K}$.

To determine the structure of the stratified space $\hat{\mathbb{R}}^3/D_2 \times Z_2^c$, we use [2, Table 31.1], where the Euler angles of each matrix of the group D_2 are given.

We calculate the matrices of all elements of the group D_2. Multiplying them by -1, we obtain the matrices for all remaining elements. It turns out that the group $D_2 \times Z_2^c$ contains 8 diagonal matrices, where each nonzero matrix entry is equal to either 1 or -1.

It is easy to see that the orbit of a point $\vec{k} \in \hat{\mathbb{R}}^3$ is an 8-point set if and only if $k_1 \neq 0, k_2 \neq 0, k_3 \neq 0$. We choose

$$(\hat{\mathbb{R}}^3/D_2 \times Z_2^c)_7 = \{\, (k_1, k_2, k_3)^\top \in \hat{\mathbb{R}}^3 \colon k_1 > 0, k_2 > 0, k_3 > 0 \,\}.$$

We have $K_7 = Z_1$.

The orbit of a point $\vec{k} \in \hat{\mathbb{R}}^3$ is a 4-point set if and only if exactly one of the numbers k_1, k_2 and k_3 is equal to 0. The strata $(\hat{\mathbb{R}}^3/D_2 \times Z_2^c)_m$, $4 \leq m \leq 6$, must be subsets of the boundary of the stratum $(\hat{\mathbb{R}}^3/D_2 \times Z_2^c)_7$. The unique possible choice up to enumeration is

$$\begin{aligned}
(\hat{\mathbb{R}}^3/D_2 \times Z_2^c)_4 &= \{\, (k_1, k_2, 0)^\top \in \hat{\mathbb{R}}^3 \colon k_1 > 0, k_2 > 0 \,\}, \\
(\hat{\mathbb{R}}^3/D_2 \times Z_2^c)_5 &= \{\, (k_1, 0, k_3)^\top \in \hat{\mathbb{R}}^3 \colon k_1 > 0, k_3 > 0 \,\}, \\
(\hat{\mathbb{R}}^3/D_2 \times Z_2^c)_6 &= \{\, (0, k_2, k_3)^\top \in \hat{\mathbb{R}}^3 \colon k_2 > 0, k_3 > 0 \,\}.
\end{aligned}$$

The stationary subgroups are $K_4 = Z_2^-(\sigma_z)$, $K_5 = Z_2^-(\sigma_y)$ and $K_6 = Z_2^-(\sigma_x)$. This notation means: K_4 is the unique subgroup of the group $\tilde{K}$ that is isomorphic to Z_2^- and contains the matrix $\sigma_z \in \tilde{K}$.

The orbit of a point $\vec{k} \in \hat{\mathbb{R}}^3$ is a 2-point set if and only if exactly two of the numbers k_1, k_2 and k_3 are equal to 0. By the same reasoning as before, the unique possible choice up to enumeration is

$$\begin{aligned}
(\hat{\mathbb{R}}^3/D_2 \times Z_2^c)_1 &= \{\, (k_1, 0, 0)^\top \in \hat{\mathbb{R}}^3 \colon k_1 > 0 \,\}, \\
(\hat{\mathbb{R}}^3/D_2 \times Z_2^c)_2 &= \{\, (0, k_2, 0)^\top \in \hat{\mathbb{R}}^3 \colon k_2 > 0 \,\}, \\
(\hat{\mathbb{R}}^3/D_2 \times Z_2^c)_3 &= \{\, (0, 0, k_3)^\top \in \hat{\mathbb{R}}^3 \colon k_3 > 0 \,\}.
\end{aligned}$$

The stationary subgroups are $K_1 = D_2^v(C_{2x})$, $K_2 = D_2^v(C_{2y})$ and $K_3 = D_2^v(C_{2z})$.

Put $\tilde{K} = D_4 \times Z_2^c$. Using [2, Table 33.1], we calculate the matrices of all elements of this group. It turns out that besides the matrices of the subgroup $D_2 \times Z_2^c$ listed before, the remaining 8 matrices are as follows: $C_4^+ = \begin{pmatrix} 0 & 1 & 0 \\ -1 & 0 & 0 \\ 0 & 0 & 1 \end{pmatrix}$, $S_4^- = -C_4^+$, $C_4^- = \begin{pmatrix} 0 & -1 & 0 \\ 1 & 0 & 0 \\ 0 & 0 & 1 \end{pmatrix}$, $S_4^+ = -C_4^-$, $C_{21}'' = \begin{pmatrix} 0 & -1 & 0 \\ -1 & 0 & 0 \\ 0 & 0 & -1 \end{pmatrix}$, $\sigma_{d1} = -C_{21}''$, $C_{22}'' = \begin{pmatrix} 0 & -1 & 0 \\ 1 & 0 & 0 \\ 0 & 0 & -1 \end{pmatrix}$, and $\sigma_{d2} = -C_{22}''$.

The orbit of a point $\vec{k} \in \hat{\mathbb{R}}^3$ is a 16-point set if and only if $k_1 \neq 0, k_2 \neq 0, k_3 \neq 0$ and $k_1 \neq k_2$. We choose

$$(\hat{\mathbb{R}}^3/D_4 \times Z_2^c)_7 = \{\, (k_1, k_2, k_3)^\top \in \hat{\mathbb{R}}^3 \colon k_1 > 0, k_2 > 0, k_3 > 0, k_1 < k_2 \,\}.$$

We have $K_7 = Z_1$.

Consider the following manifold lying in the boundary of $(\hat{\mathbb{R}}^3/D_4 \times Z_2^c)_7$.

$$(\hat{\mathbb{R}}^3/D_4 \times Z_2^c)_6 = \{\, (k_1, k_2, 0)^\top \in \hat{\mathbb{R}}^3 \colon 0 < k_1 < k_2 \,\}\,.$$

The stationary subgroup of this stratum is $K_6 = Z_2^-(\sigma_h)$.

The next two-dimensional stratum lying in the boundary of $(\hat{\mathbb{R}}^3/D_4 \times Z_2^c)_7$ is

$$(\hat{\mathbb{R}}^3/D_4 \times Z_2^c)_5 = \{\, (0, k_2, k_3)^\top \in \hat{\mathbb{R}}^3 \colon k_2 > 0, k_3 > 0 \,\}\,.$$

The stationary subgroup of this stratum is $K_5 = Z_2^-(\sigma_{v1})$.

The last two-dimensional stratum lying in the boundary of $(\hat{\mathbb{R}}^3/D_4 \times Z_2^c)_7$ is

$$(\hat{\mathbb{R}}^3/D_4 \times Z_2^c)_4 = \{\, (k_1, k_2, k_3)^\top \in \hat{\mathbb{R}}^3 \colon k_1 = k_2 > 0, k_3 > 0 \,\}\,.$$

The stationary subgroup of this stratum is $K_4 = Z_2^-(\sigma_{d1})$.

The one-dimensional stratum lying in the common boundary of $(\hat{\mathbb{R}}^3/D_4 \times Z_2^c)_7$, $(\hat{\mathbb{R}}^3/D_4 \times Z_2^c)_6$ and $(\hat{\mathbb{R}}^3/D_4 \times Z_2^c)_5$ is

$$(\hat{\mathbb{R}}^3/D_4 \times Z_2^c)_3 = \{\, (0, k_2, 0)^\top \in \hat{\mathbb{R}}^3 \colon k_2 > 0 \,\}\,.$$

The stationary subgroup is $K_3 = D_2^v(C'_{22}, \sigma_{v1}, \sigma_h)$.

The one-dimensional stratum lying in the common boundary of $(\hat{\mathbb{R}}^3/D_4 \times Z_2^c)_7$, $(\hat{\mathbb{R}}^3/D_4 \times Z_2^c)_6$ and $(\hat{\mathbb{R}}^3/D_4 \times Z_2^c)_4$ is

$$(\hat{\mathbb{R}}^3/D_4 \times Z_2^c)_2 = \{\, (k_1, k_2, 0)^\top \in \hat{\mathbb{R}}^3 \colon k_1 = k_2 > 0 \,\}\,.$$

The stationary subgroup is $K_2 = D_2^v(C''_{22}, \sigma_{d1}, \sigma_h)$.

Finally, the one-dimensional stratum lying in the common boundary of $(\hat{\mathbb{R}}^3/D_4 \times Z_2^c)_7$, $(\hat{\mathbb{R}}^3/D_4 \times Z_2^c)_5$ and $(\hat{\mathbb{R}}^3/D_4 \times Z_2^c)_4$ is

$$(\hat{\mathbb{R}}^3/D_4 \times Z_2^c)_1 = \{\, (0, 0, k_3)^\top \in \hat{\mathbb{R}}^3 \colon k_3 > 0 \,\}\,.$$

The stationary subgroup is $K_1 = D_4^v$.

Put $\tilde{K} = \mathcal{T} \times Z_2^c$. Using [2, Table 71.1], we calculate the matrices of all elements of this group. It turns out that besides the matrices of the group $D_2 \times Z_2^c$ calculated above, the remaining 16 matrices are as follows: $C_{31}^+ = \begin{pmatrix} 0 & 0 & 1 \\ -1 & 0 & 0 \\ 0 & -1 & 0 \end{pmatrix}$, $S_{61}^- = -C_{31}^+$, $C_{32}^+ = \begin{pmatrix} 0 & 0 & -1 \\ -1 & 0 & 0 \\ 0 & 1 & 0 \end{pmatrix}$, $S_{62}^- = -C_{32}^+$, $C_{33}^+ = \begin{pmatrix} 0 & 0 & -1 \\ 1 & 0 & 0 \\ 0 & -1 & 0 \end{pmatrix}$, $S_{63}^- = -C_{33}^+$, $C_{34}^+ = \begin{pmatrix} 0 & 0 & 1 \\ 1 & 0 & 0 \\ 0 & 1 & 0 \end{pmatrix}$, $S_{64}^- = -C_{34}^+$, $C_{31}^- = \begin{pmatrix} 0 & -1 & 0 \\ 0 & 0 & -1 \\ 1 & 0 & 0 \end{pmatrix}$, $S_{61}^+ = -C_{31}^-$, $C_{32}^- = \begin{pmatrix} 0 & -1 & 0 \\ 0 & 0 & 1 \\ -1 & 0 & 0 \end{pmatrix}$, $S_{62}^+ = -C_{32}^-$, $C_{33}^- = \begin{pmatrix} 0 & 1 & 0 \\ 0 & 0 & -1 \\ -1 & 0 & 0 \end{pmatrix}$, $S_{63}^+ = -C_{33}^-$, $C_{34}^- = \begin{pmatrix} 0 & 1 & 0 \\ 0 & 0 & 1 \\ 1 & 0 & 0 \end{pmatrix}$, $S_{64}^+ = -C_{34}^-$.

Consider the stratum

$$(\hat{\mathbb{R}}^3/\mathcal{T} \times Z_2^c)_5 = \{\, (k_1, k_2, k_3)^\top \in \hat{\mathbb{R}}^3 \colon 0 < k_3 < \min\{k_1, k_2\}\,\}\,.$$

The union of the orbits of the points of this set under the action of the group $\mathcal{T} \times Z_2^c$ is the set of points $(k_1, k_2, k_3)^\top \in \hat{\mathbb{R}}^3$ such that $k_1 \neq 0$, $k_2 \neq 0$, $k_3 \neq 0$, and two of the three numbers $|k_1|$, $|k_2|$ and $|k_3|$ are greater than the remaining one. The above union contains 24 connected components and is dense in $\hat{\mathbb{R}}^3$. The orbit of a point $\vec{k} \in (\hat{\mathbb{R}}^3/\mathcal{T} \times Z_2^c)_5$ is a 24-point set, in particular, its stationary subgroup is $K_5 = Z_1$. As we will see, there are more points with a 24-point orbit!

Indeed, consider the following stratum in the boundary of the set $(\hat{\mathbb{R}}^3/\mathcal{T} \times Z_2^c)_5$:

$$(\hat{\mathbb{R}}^3/\mathcal{T} \times Z_2^c)_4 = \{\, (k_1, k_2, k_3)^\top \in \hat{\mathbb{R}}^3 \colon 0 < k_3 = k_2 < k_1\,\}\,.$$

The stationary subgroup of this stratum is $K_4 = Z_1$. In particular, the matrix C_{34}^+ maps it to the set $\{\, (k_1, k_2, k_3)^\top \in \hat{\mathbb{R}}^3 \colon 0 < k_3 = k_1 < k_2\,\}$, which is another part of the boundary of the stratum $(\hat{\mathbb{R}}^3/\mathcal{T} \times Z_2^c)_5$, thus the subset $\hat{\mathbb{R}}^3/\mathcal{T} \times Z_2^c$ is not closed in $\hat{\mathbb{R}}^3$!

The points of the stratum

$$(\hat{\mathbb{R}}^3/\mathcal{T} \times Z_2^c)_3 = \{\, (k_1, k_2, 0)^\top \in \hat{\mathbb{R}}^3 \colon k_1 > 0, k_2 > 0\,\}$$

have 12-point orbits, and the stationary subgroup is $K_3 = Z_2^-(\sigma_z)$.

The points of the one-dimensional stratum

$$(\hat{\mathbb{R}}^3/\mathcal{T} \times Z_2^c)_2 = \{\, (k_1, k_2, k_3)^\top \in \hat{\mathbb{R}}^3 \colon 0 < k_1 = k_2 = k_3\,\}$$

have 8-point orbits. The stationary subgroup is $K_2 = Z_3$.

Finally, the points of the stratum

$$(\hat{\mathbb{R}}^3/\mathcal{T} \times Z_2^c)_1 = \{\, (k_1, 0, 0)^\top \in \hat{\mathbb{R}}^3 \colon k_1 > 0\,\}$$

have 6-point orbits. In particular, any of the matrices S_{61}^-, S_{62}^-, C_{33}^+, or C_{34}^+, map the stratum $(\hat{\mathbb{R}}^3/\mathcal{T} \times Z_2^c)_1$ to the set $\{\, (0, k_2, 0)^\top \in \hat{\mathbb{R}}^3 \colon k_2 > 0\,\}$, another part of the boundary of the stratum $(\hat{\mathbb{R}}^3/\mathcal{T} \times Z_2^c)_5$. The stationary subgroup is $K_1 = D_2^v$.

Put $\tilde{K} = O \times Z_2^c$. Using [2, Table 71.1], we calculate the matrices of all elements of this group. It turns out that the 48 matrices of the group $O \times Z_2^c$ have exactly one nonzero matrix entry in each row and in each column, and the nonzero entries may take values ± 1.

It is easy to check that the orbit of a point $\vec{k} \in \hat{\mathbb{R}}^3$ is a 48-point set if and only if the set $\{|k_1|, |k_2|, |k_3|\}$ contains 3 nonzero elements. We choose

$$(\hat{\mathbb{R}}^3/O \times Z_2^c)_7 = \{\, (k_1, k_2, k_3)^\top \in \hat{\mathbb{R}}^3 \colon 0 < k_1 < k_2 < k_3\,\}\,.$$

The stationary subgroup of this stratum is $K_7 = Z_1$.

Table 4.1 The stationary subgroups of the strata

	$\tilde{K}$			
	$D_2 \times Z_2^c$	$D_4 \times Z_2^c$	$\mathcal{T} \times Z_2^c$	$\mathcal{O} \times Z_2^c$
K_1	$D_2^v(C_{2x})$	D_4^v	D_2^v	D_v^4
K_2	$D_2^v(C_{2y})$	$D_2^v(C''_{22}, \sigma_{d1}, \sigma_h)$	Z_3	D_v^3
K_3	$D_2^v(C_{2z})$	$D_2^v(C'_{22}, \sigma_{v1}, \sigma_h)$	$Z_2^-(\sigma_z)$	$D_2^v(C'_{2f}, \sigma_{d4}, \sigma_x)$
K_4	$Z_2^-(\sigma_z)$	$Z_2^-(\sigma_{d1})$	Z_1	$Z_2^-(\sigma_{d4})$
K_5	$Z_2^-(\sigma_y)$	$Z_2^-(\sigma_{v1})$	Z_1	$Z_2^-(\sigma_{d1})$
K_6	$Z_2^-(\sigma_x)$	$Z_2^-(\sigma_h)$	–	$Z_2^-(\sigma_x)$
K_7	Z_1	Z_1	–	Z_1

The three two-dimensional strata that are subsets of the boundary of the stratum $(\hat{\mathbb{R}}^3/\mathcal{O} \times Z_2^c)_7$ are as follows:

$$(\hat{\mathbb{R}}^3/\mathcal{O} \times Z_2^c)_6 = \{\, (k_1, k_2, k_3)^\top \in \hat{\mathbb{R}}^3 \colon 0 = k_1 < k_2 < k_3 \,\},$$
$$(\hat{\mathbb{R}}^3/\mathcal{O} \times Z_2^c)_5 = \{\, (k_1, k_2, k_3)^\top \in \hat{\mathbb{R}}^3 \colon 0 < k_1 = k_2 < k_3 \,\},$$
$$(\hat{\mathbb{R}}^3/\mathcal{O} \times Z_2^c)_4 = \{\, (k_1, k_2, k_3)^\top \in \hat{\mathbb{R}}^3 \colon 0 < k_1 < k_2 = k_3 \,\}.$$

Their stationary subgroups are $K_6 = Z_2^-(\sigma_x)$, $K_5 = Z_2^-(\sigma_{d1})$ and $K_4 = Z_2^-(\sigma_{d4})$.

The three one-dimensional strata that are subsets of the boundary of the stratum $(\hat{\mathbb{R}}^3/\mathcal{O} \times Z_2^c)_7$ are as follows:

$$(\hat{\mathbb{R}}^3/\mathcal{O} \times Z_2^c)_3 = \{\, (k_1, k_2, k_3)^\top \in \hat{\mathbb{R}}^3 \colon 0 = k_1 < k_2 = k_3 \,\},$$
$$(\hat{\mathbb{R}}^3/\mathcal{O} \times Z_2^c)_2 = \{\, (k_1, k_2, k_3)^\top \in \hat{\mathbb{R}}^3 \colon 0 < k_1 = k_2 = k_3 \,\},$$
$$(\hat{\mathbb{R}}^3/\mathcal{O} \times Z_2^c)_1 = \{\, (k_1, k_2, k_3)^\top \in \hat{\mathbb{R}}^3 \colon 0 = k_1 = k_2 < k_3 \,\}.$$

Their stationary subgroups are $K_3 = D_2^v(C'_{2f}, \sigma_{d4}, \sigma_x)$, $K_2 = D_3^v$, and $K_1 = D_4^v$.

We summarise our findings in Table 4.1.

4.5 Invariant Subspaces

In this section, we begin to analyse the second condition in (4.18).

Consider the restriction of the function $f(\vec{k})$ to the stratum $(\hat{\mathbb{R}}^3/\tilde{K})_m$. By definition of the stationary subgroup, for a point $\vec{\lambda}_m \in (\hat{\mathbb{R}}^3/\tilde{K})_m$ we have

$$\tilde{g}\vec{\lambda}_m = \vec{\lambda}_m, \qquad \tilde{g} \in K_m\,.$$

Substitute this equation into the second condition in (4.18). We obtain

$$f(\vec{\lambda}_m) = \tilde{g}\vec{\lambda}_m, \qquad \tilde{g} \in K_m\,. \tag{4.20}$$

Denote by $\tilde{\theta}_m$ the restriction of the representation $\tilde{\theta}$ to the subgroup K_m. Equation (4.20) means that $f(\vec{\lambda}_m)$ belongs to the invariant subspace $\tilde{\mathsf{V}}_m$, where the representation $\tilde{\theta}_m$ acts trivially. In the following example, we find the basis tensors of the spaces $\tilde{\mathsf{V}}_m$.

Example 4.4 Let $G = Z_1$, $K = \mathrm{O}(3)$. Put $m = 0$. Example 4.3 gives $K_0 = \mathrm{O}(3)$. By Eq. (4.6), $\dim \tilde{\mathsf{V}}_0 = 5$. By definition of the Godunov–Gordienko coefficients, we obtain the basis of the space $\tilde{\mathsf{V}}_0$ in the form

$$\begin{aligned}
\mathsf{f}^1 &= \frac{1}{\sqrt{3}}(\mathsf{d}^1 \otimes \mathsf{d}^1 + \mathsf{d}^2 \otimes \mathsf{d}^2 + \mathsf{d}^3 \otimes \mathsf{d}^3)\,,\\
\mathsf{f}^2 &= \frac{1}{\sqrt{3}}(\mathsf{d}^4 \otimes \mathsf{d}^4 + \mathsf{d}^5 \otimes \mathsf{d}^5 + \mathsf{d}^6 \otimes \mathsf{d}^6)\,,\\
\mathsf{f}^3 &= \frac{1}{\sqrt{5}}(\mathsf{d}^7 \otimes \mathsf{d}^7 + \cdots + \mathsf{d}^{11} \otimes \mathsf{d}^{11})\,,\\
\mathsf{f}^4 &= \frac{1}{\sqrt{7}}(\mathsf{d}^{12} \otimes \mathsf{d}^{12} + \cdots + \mathsf{d}^{18} \otimes \mathsf{d}^{18})\,,\\
\mathsf{f}^5 &= \frac{1}{\sqrt{6}}(\mathsf{d}^1 \otimes \mathsf{d}^4 + \mathsf{d}^4 \otimes \mathsf{d}^1 + \mathsf{d}^2 \otimes \mathsf{d}^5 + \mathsf{d}^5 \otimes \mathsf{d}^2 + \mathsf{d}^3 \otimes \mathsf{d}^6 + \mathsf{d}^6 \otimes \mathsf{d}^3)\,,
\end{aligned} \tag{4.21}$$

where d^1, …, d^{18} are the basis tensors (3.2), (3.3), (3.4), and (3.5).

Put $m = 1$. Example 4.3 gives $K_1 = \mathrm{O}(2)$. It is known that the restriction of the representation θ^ℓ of the group O(3) to the subgroup O(2) contains a copy of the trivial representation of O(2) if and only if ℓ is even. Equation (4.6) gives

$$\dim \tilde{\mathsf{V}}_1 = 5 + 10 + 5 + 1 = 21\,.$$

In addition to the tensors (4.21), we have, for example:

$$\mathsf{f}^6 = \sum_{i,j=-1}^{1} g^{0[i,j]}_{2[1,1]}\mathsf{d}^{i+2} \otimes \mathsf{d}^{j+2} = -\frac{1}{\sqrt{6}}\mathsf{d}^1 \otimes \mathsf{d}^1 + \frac{\sqrt{2}}{\sqrt{3}}\mathsf{d}^2 \otimes \mathsf{d}^2 - \frac{1}{\sqrt{6}}\mathsf{d}^3 \otimes \mathsf{d}^3\,,$$

and so on. Calculation of the tensors f^7, …, f^{21} may be left to the reader. See also [20].

In the remaining part of this example, $G = D_2$, and we use Table 4.1. Let $K = D_2$. Equation (4.11) gives $\tilde{\theta} = 6A_g \oplus 3A_u$. If $m = 0$, we have $K_0 = D_2 \times Z^c_2$. The invariant subspace $\tilde{\mathsf{V}}_0$ is the linear space of all symmetric 3×3 matrices with real matrix entries. We choose the following basis of this space:

$$\mathsf{f}^{A_g,i1} = \mathsf{d}^i \otimes \mathsf{d}^i, \qquad 1 \le i \le 3, \tag{4.22a}$$

$$\mathsf{f}^{A_g,41} = \frac{1}{\sqrt{2}}(\mathsf{d}^2 \otimes \mathsf{d}^3 + \mathsf{d}^3 \otimes \mathsf{d}^2), \tag{4.22b}$$

$$\mathsf{f}^{A_g,51} = \frac{1}{\sqrt{2}}(\mathsf{d}^1 \otimes \mathsf{d}^3 + \mathsf{d}^3 \otimes \mathsf{d}^1), \tag{4.22c}$$

$$\mathsf{f}^{A_g,61} = \frac{1}{\sqrt{2}}(\mathsf{d}^1 \otimes \mathsf{d}^2 + \mathsf{d}^2 \otimes \mathsf{d}^1), \tag{4.22d}$$

where d^i are the tensors of the basis (3.6).

The same situation occurs when the subgroup K_m is not a subgroup of the group SO(3), that is, for K_1–K_6. We have $\tilde{\mathsf{V}}_m = \tilde{\mathsf{V}}_0$ for $1 \le m \le 6$. On the other hand, $K_7 = Z_1$, and $\tilde{\mathsf{V}}_7$ is the linear space of all Hermitian 3×3 matrices. In addition to the basis tensors of the subspace $\tilde{\mathsf{V}}_0$ described above, three new basis tensors appear:

$$\mathsf{f}^{A_u,11} = \frac{\mathrm{i}}{\sqrt{2}}(\mathsf{d}^2 \otimes \mathsf{d}^3 - \mathsf{d}^3 \otimes \mathsf{d}^2), \tag{4.23a}$$

$$\mathsf{f}^{A_u,21} = \frac{\mathrm{i}}{\sqrt{2}}(\mathsf{d}^1 \otimes \mathsf{d}^3 - \mathsf{d}^3 \otimes \mathsf{d}^1), \tag{4.23b}$$

$$\mathsf{f}^{A_u,31} = \frac{\mathrm{i}}{\sqrt{2}}(\mathsf{d}^1 \otimes \mathsf{d}^2 - \mathsf{d}^2 \otimes \mathsf{d}^1). \tag{4.23c}$$

Put $K = D_2 \times Z_2^c$. Equation (4.7) gives $\tilde{\theta} = 6A_g$. For any m, $0 \le m \le 7$, the restriction of the trivial representation $6A_g$ to the subgroup K_m is itself trivial. The invariant subspace $\tilde{\mathsf{V}}_m$ is the linear space of all symmetric 3×3 matrices with real matrix entries, and we choose the basis tensors constructed above.

Put $K = D_4$. Equation (4.12) gives $\tilde{\theta} = 4A_{1g} \oplus 2B_{1g} \oplus A_{1u} \oplus 2B_{1u}$. The 4 copies of the trivial representation A_{1g} act on the one-dimensional spaces generated by the tensors (4.22a) and (4.22b), where d^i are the tensors of the basis (3.7). The two copies of the representation B_{1g} act on the one-dimensional spaces generated by the tensors (4.22c) and (4.22d). The representation A_{1u} acts on the one-dimensional space generated by the tensor (4.23a). The two copies of the representation B_{1u} act on the one-dimensional spaces generated by the tensors (4.23b) and (4.23c).

The restriction of the representation $\tilde{\theta}$ to the subgroup $K_0 = D_4 \times Z_2^c$ contains 4 copies of the trivial representation A_{1g}. Then, $\tilde{\mathsf{V}}_0$ is the 4-dimensional space generated by the tensors (4.22a) and (4.22b). By [2, Table 33.9], the same holds true for $K_1 = D_4^v$, $K_2 = D_2^v(C''_{22}, \sigma_{d1}, \sigma_h)$, $K_5 = Z_2^-(\sigma_{v1})$ and $K_6 = Z_2^-(\sigma_h)$. We have $\tilde{\mathsf{V}}_m = \tilde{\mathsf{V}}_0$, $m = 1, 2, 5, 6$.

The restriction of $\tilde{\theta}$ to the subgroup $K_3 = D_2^v(C'_{22}, \sigma_{v1}, \sigma_h)$ becomes trivial for the components $4A_{1g}$ and $2B_{1g}$. The space $\tilde{\mathsf{V}}_3$ is the 6-dimensional space generated by the tensors (4.22).

The restriction of $\tilde{\theta}$ to the subgroup $K_4 = Z_2^-(\sigma_{d1})$ becomes trivial for the components $4A_{1g}$ and $2B_{1u}$. The space $\tilde{\mathsf{V}}_4$ is the 6-dimensional space generated by the tensors (4.22a), (4.22b), (4.23b), and (4.23c).

Finally, the restriction of $\tilde{\theta}$ to the subgroup $K_7 = Z_1$ is trivial. The 9-dimensional space $\tilde{\mathsf{V}}_7$ is generated by the tensors (4.22) and (4.23).

Put $K = D_4^h$. Equation (4.15) gives $\tilde{\theta} = 4A_{1g} \oplus 2B_{1g} \oplus A_{2u} \oplus 2B_{2u}$. Similarly to the case of $K = D_4$, the 4 copies of the trivial representation A_{1g} act on the one-dimensional spaces generated by the tensors (4.22a) and (4.22b), where d^i are the tensors of the basis (3.7). The two copies of the representation B_{1g} act on the one-dimensional spaces generated by the tensors (4.22c) and (4.22d). The representation A_{2u} acts on the one-dimensional space generated by the tensor (4.23a). The two copies of the representation B_{2u} act on the one-dimensional spaces generated by the tensors (4.23b) and (4.23c).

We have: $\tilde{\mathsf{V}}_0$ is the 4-dimensional space generated by the tensors (4.22a) and (4.22b) by the same reasoning as before. By [2, Table 33.9], the same holds true for $K_2 = D_2^v(C_{22}'', \sigma_{d1}, \sigma_h)$. We have $\tilde{\mathsf{V}}_2 = \tilde{\mathsf{V}}_0$.

The restriction of $\tilde{\theta}$ to the subgroups $K_1 = D_4^v$ and $K_4 = Z_2^-(\sigma_{d1})$ becomes trivial for the components $4A_{1g}$ and A_{2u}. The spaces $\tilde{\mathsf{V}}_1$ and $\tilde{\mathsf{V}}_4$ are 5-dimensional. They are generated by the tensors (4.22a), (4.22b) and (4.23a).

The restriction of $\tilde{\theta}$ to the subgroups $K_3 = D_2^v(C_{22}', \sigma_{v1}, \sigma_h)$ and $K_6 = Z_2^-(\sigma_h)$ becomes trivial for the components $4A_{1g}$ and $2B_{1g}$. The spaces $\tilde{\mathsf{V}}_3$ and $\tilde{\mathsf{V}}_6$ are 6-dimensional. They are generated by the tensors (4.22).

Finally, the restriction of $\tilde{\theta}$ to the subgroups $K_5 = Z_2^-(\sigma_{v1})$ and $K_7 = Z_1$ is trivial. The spaces $\tilde{\mathsf{V}}_5$ and $\tilde{\mathsf{V}}_7$ are 9-dimensional. They are generated by the tensors (4.22) and (4.23).

Put $K = D_4 \times Z_2^c$. Equation (4.8) gives $\tilde{\theta} = 3A_{1g} \oplus A_{2g} \oplus B_{1g} \oplus B_{2g}$. The 3 copies of the trivial representation A_{1g} act on the one-dimensional spaces generated by the tensors (4.22a), where d^i are the tensors of the basis (3.7). The representation A_{2g} acts on the one-dimensional space generated by the tensor (4.22b). The representation B_{2g} acts on the one-dimensional space generated by the tensor (4.22c). The representation B_{1g} acts on the one-dimensional space generated by the tensor (4.22d).

$\tilde{\mathsf{V}}_0$ is the 3-dimensional space generated by the tensors (4.22a) by the same reasoning as before. By [2, Table 33.9], the same holds true for $K_1 = D_4^v$. We have $\tilde{\mathsf{V}}_1 = \tilde{\mathsf{V}}_0$.

The restriction of $\tilde{\theta}$ to the subgroups $K_2 = D_2^v(C_{22}'', \sigma_{d1}, \sigma_h)$ and $K_4 = Z_2^-(\sigma_{d1})$ becomes trivial for the components $3A_{1g}$ and B_{2g}. The spaces $\tilde{\mathsf{V}}_2$ and $\tilde{\mathsf{V}}_4$ are 4-dimensional. They are generated by the tensors (4.22a) and (4.22c).

The restriction of $\tilde{\theta}$ to the subgroups $K_3 = D_2^v(C_{22}', \sigma_{v1}, \sigma_h)$ and $K_5 = Z_2^-(\sigma_{v1})$ becomes trivial for the components $3A_{1g}$ and B_{1g}. The spaces $\tilde{\mathsf{V}}_3$ and $\tilde{\mathsf{V}}_5$ are 4-dimensional. They are generated by the tensors (4.22a) and (4.22d).

Finally, the restriction of $\tilde{\theta}$ to the subgroups $K_6 = Z_2^-(\sigma_h)$ and $K_7 = Z_1$ is trivial. The spaces $\tilde{\mathsf{V}}_6$ and $\tilde{\mathsf{V}}_7$ are 6-dimensional. They are generated by the tensors (4.22).

Put $K = \mathcal{T}$. Equation (4.13) gives $\tilde{\theta} = 2A_g \oplus 2E_g \oplus A_u \oplus E_u$. The 2 copies of the trivial representation A_g act on the one-dimensional spaces generated by the tensors

$$\mathsf{f}^{A_g,11} = \mathsf{d}^1 \otimes \mathsf{d}^1, \qquad \mathsf{f}^{A_g,21} = \frac{1}{\sqrt{2}}(\mathsf{d}^2 \otimes \mathsf{d}^2 + \mathsf{d}^3 \otimes \mathsf{d}^3), \tag{4.24}$$

where d^i are the tensors given by (3.8). The first copy of the representation E_g acts on the 2-dimensional space generated by the tensors

$$\mathsf{f}^{E_g,11} = \frac{1}{\sqrt{2}}(-\mathsf{d}^2 \otimes \mathsf{d}^2 + \mathsf{d}^3 \otimes \mathsf{d}^3), \qquad \mathsf{f}^{E_g,12} = \frac{1}{\sqrt{2}}(\mathsf{d}^2 \otimes \mathsf{d}^3 + \mathsf{d}^3 \otimes \mathsf{d}^2). \tag{4.25}$$

The second copy of E_g acts on the 2-dimensional space generated by the tensors

$$\mathsf{f}^{E_g,21} = \frac{1}{\sqrt{2}}(\mathsf{d}^1 \otimes \mathsf{d}^2 + \mathsf{d}^2 \otimes \mathsf{d}^1), \qquad \mathsf{f}^{E_g,22} = \frac{1}{\sqrt{2}}(\mathsf{d}^1 \otimes \mathsf{d}^3 + \mathsf{d}^3 \otimes \mathsf{d}^1). \tag{4.26}$$

The representation A_u acts on the one-dimensional space generated by the tensor

$$\mathsf{f}^{A_u,11} = \frac{\mathrm{i}}{\sqrt{2}}(\mathsf{d}^2 \otimes \mathsf{d}^3 - \mathsf{d}^3 \otimes \mathsf{d}^2). \tag{4.27}$$

The representation E_u acts on the 2-dimensional space generated by the tensors

$$\mathsf{f}^{E_u,11} = \frac{\mathrm{i}}{\sqrt{2}}(\mathsf{d}^1 \otimes \mathsf{d}^2 - \mathsf{d}^2 \otimes \mathsf{d}^1), \qquad \mathsf{f}^{E_u,12} = \frac{\mathrm{i}}{\sqrt{2}}(\mathsf{d}^1 \otimes \mathsf{d}^3 - \mathsf{d}^3 \otimes \mathsf{d}^1). \tag{4.28}$$

$\tilde{\mathsf{V}}_0$ is the 2-dimensional space generated by the tensors (4.24) by the same reasoning as before.

By [2, Table 72.9], the restriction of $\tilde{\theta}$ to the subgroups $K_1 = D_2^y$ and $K_3 = Z_2^-(\sigma_z)$ becomes trivial for the components $2A_g$ and $2E_g$. The spaces $\tilde{\mathsf{V}}_1$ and $\tilde{\mathsf{V}}_3$ are 6-dimensional. They are generated by the tensors (4.24)–(4.26).

The restriction of $\tilde{\theta}$ to the subgroup $K_2 = Z_3$ becomes trivial for the components $2A_g$ and A_u. The space $\tilde{\mathsf{V}}_2$ is 3-dimensional. It is generated by the tensors (4.24) and (4.27). The same space appears when $K = \mathcal{O}$ and $m = 2$.

Finally, the restriction of $\tilde{\theta}$ to the subgroups $K_4 = K_5 = Z_1$ is trivial. The spaces $\tilde{\mathsf{V}}_4$ and $\tilde{\mathsf{V}}_5$ are 9-dimensional. They are generated by the tensors (4.24)–(4.28).

Put $K = \mathcal{T} \times Z_2^c$. Equation (4.9) gives $\tilde{\theta} = 2A_g \oplus 2E_g$. $\tilde{\mathsf{V}}_0$ is the 2-dimensional space generated by the tensors (4.24) by the same reasoning as before. By [2, Table 72.9], the same holds true for $K_2 = Z_3$. We have $\tilde{\mathsf{V}}_2 = \tilde{\mathsf{V}}_0$. The spaces $\tilde{\mathsf{V}}_1$ and $\tilde{\mathsf{V}}_3$ are the same as in the case of $K = \mathcal{T}$.

The restriction of $\tilde{\theta}$ to the subgroups $K_4 = K_5 = Z_1$ is trivial. The spaces $\tilde{\mathsf{V}}_4$ and $\tilde{\mathsf{V}}_5$ are 6-dimensional. They are generated by the tensors (4.24)–(4.26).

Put $K = \mathcal{O}$. Equation (4.14) gives $\tilde{\theta} = 2A_{1g} \oplus 2E_g \oplus A_{2u} \oplus E_u$. The 2 copies of the trivial representation A_{1g} act on the one-dimensional spaces generated by the tensors (4.24). The first copy of the representation E_g acts on the 2-dimensional space generated by the tensors (4.25). The second copy of E_g acts on the 2-dimensional space generated by the tensors (4.26). The representation A_{2u} acts on the one-dimensional space generated by the tensor (4.27). The representation E_u acts on the 2-dimensional space generated by the tensors (4.28).

The space $\tilde{\mathsf{V}}_0$ is the 2-dimensional space generated by the tensors (4.24), because the restriction of the trivial representation A_{1g} to any subgroup is trivial.

A new problem arises when we determine the basis of the space $\tilde{\mathsf{V}}_1$. By [2, Table 71.9], the restriction of the representation A_{1g} to the subgroup $K_1 = D_4^v$ is trivial, and the above basis contains the tensors (4.24). The restrictions of the representations A_{2u} and E_u to K_1 do not have trivial components. The restriction of the representation E_g to K_1 is equivalent to $A_1 \oplus B_1$. On which space does the trivial component A_1 act?

To solve this problem, we find the matrices of the representation E_g for all 8 elements of the subgroup K_1. A common eigenvector of all of them corresponding to the eigenvalue 1 generates the one-dimensional subspace on which the trivial component A_1 acts.

Unfortunately, [2, Table 69.7] contains the matrices of the *unitary* representation E_g. To overcome this difficulty, we consider [2, Table 71.5]. According to this table, the restriction of the representation θ^1 of the group O(3) to the subgroup $O \times Z_2^c$ is T_1. The matrix entries g_{ij} of the *orthogonal* representation T_1 in the standard basis $\{\vec{e}^i : 1 \leq i \leq 3\}$ have already been calculated (the 48 matrices mentioned before). By the above table, the tensor square $T_1 \otimes T_1$ contains E_g, and the basis in the space of this component is given by

$$\mathsf{f}^i = \sum_{j,k=1}^{3} c^{i[j,k]}_{E_g[T_1,T_1]} \vec{e}^j \otimes \vec{e}^k ,$$

where the nonzero Clebsch–Gordan coefficients $c^{i[j,k]}_{E[T_1,T_1]}$ are as follows: $c^{1[1,1]}_{E_g[T_1,T_1]} = \frac{1}{\sqrt{2}}$, $c^{1[2,2]}_{E_g[T_1,T_1]} = -\frac{1}{\sqrt{2}}$, $c^{2[1,1]}_{E_g[T_1,T_1]} = c^{2[2,2]}_{E_g[T_1,T_1]} = -\frac{1}{\sqrt{6}}$, and $c^{2[3,3]}_{E_g[T_1,T_1]} = \frac{\sqrt{2}}{\sqrt{3}}$. In this basis, the matrix entries of the representation E_g become

$$(E_g(g))_{ij} = \sum_{k,l,m,n=1}^{3} c^{i[k,l]}_{E_g[T_1,T_1]} (T_1(g))_{km} (T_1(g))_{ln} c^{j[m,n]}_{E_g[T_1,T_1]} .$$

Calculations give the following result. The representation E_g maps the matrices E, C_2, σ_{v1} and σ_{d2} to the 2×2 identity matrix, and the remaining elements of the subgroup D_4^v to the matrix $\left(\begin{smallmatrix}-1 & 0\\ 0 & 1\end{smallmatrix}\right)$. The trivial components A_1 act on the one-dimensional spaces generated by the tensors $\mathsf{f}^{E_g,12}$ of (4.25) and $\mathsf{f}^{E_g,22}$ of (4.26). The space $\tilde{\mathsf{V}}_1$ is 4-dimensional.

Consider the space $\tilde{\mathsf{V}}_2$. By [2, Tables 71.9, 73.9], the restriction of the representations E_g and E_u to the subgroup $K_2 = D_3^v$ is nontrivial, while the restriction of the representations A_{1g} and A_{2u} to the same subgroup is trivial. The space $\tilde{\mathsf{V}}_2$ is 3-dimensional, and the basis is (4.24) and (4.27).

Consider the space $\tilde{\mathsf{V}}_3$. By [2, Table 71.4], the restriction of A_{2u} to $K_3 = D_2^{\nu}$ is nontrivial. The representation E_g maps the elements E and σ_x of the subgroup D_2^{ν} to the identity matrix, while the elements C'_{2f} and σ_{d4} are mapped to the matrix $\frac{1}{2}\begin{pmatrix} 1 & -\sqrt{3} \\ -\sqrt{3} & -1 \end{pmatrix}$. The common eigenvector of the above matrices with unit norm corresponding to the eigenvalue 1 is $\frac{1}{2}(1, -\sqrt{3})^{\top}$. The basis tensors of the subspaces on which A_1 acts are

$$\mathsf{f}^{3'} = \frac{1}{2}(\mathsf{f}^{E_g,11} - \sqrt{3}\mathsf{f}^{E_g,12}), \qquad \mathsf{f}^{4'} = \frac{1}{2}(\mathsf{f}^{E_g,21} - \sqrt{3}\mathsf{f}^{E_g,22}), \tag{4.29}$$

where $\mathsf{f}^{E_g,11}$ and $\mathsf{f}^{E_g,12}$ are given by (4.25), and $\mathsf{f}^{E_g,21}$ and $\mathsf{f}^{E_g,22}$ are given by (4.26).

The representation E_u maps the element E to the identity matrix, and the element σ_x to the negative identity matrix. None of the eigenvalues of $E_u(\sigma_x)$ are equal to 1. The representation E_u does not contain trivial components. The space $\tilde{\mathsf{V}}_3$ is 4-dimensional, and the basis is (4.24) and (4.29).

Consider the space $\tilde{\mathsf{V}}_4$. By [2, Table 71.4], the restriction of A_{2u} to $K_4 = Z_2^-(\sigma_{d4})$ is trivial. The matrix $E_g(\sigma_{d4})$ has already been calculated, and the basis tensors are (4.29). We have $E_u(\sigma_{d4}) = -E_g(\sigma_{d4})$. No eigenvalues of this matrix are equal to 1. The space $\tilde{\mathsf{V}}_4$ is 5-dimensional, and the basis is (4.24), (4.27), and (4.29).

Consider the space $\tilde{\mathsf{V}}_5$. By [2, Table 71.4], the restriction of A_{2u} to $K_5 = Z_2^-(\sigma_{d1})$ is trivial. The matrix $E_g(\sigma_{d1})$ has already been calculated. The trivial components A_1 again act on the one-dimensional spaces generated by the tensors $\mathsf{f}^{E_g,12}$ of (4.25) and $\mathsf{f}^{E_g,22}$ of (4.26). We have $E_u(\sigma_{d1}) = -E_g(\sigma_{d1})$. The trivial component A_1 acts on the space generated by the tensor $\mathsf{f}^{E_u,11}$ of (4.28). The space $\tilde{\mathsf{V}}_5$ is 6-dimensional.

Consider the space $\tilde{\mathsf{V}}_6$. By [2, Table 71.4], the restriction of A_{2u} to $K_6 = Z_2^-(\sigma_x)$ is nontrivial. The matrix $E_g(\sigma_x)$ has been calculated before and is equal to the identity matrix. The four copies of the trivial representation A_1 act on the two 2-dimensional spaces with bases (4.25) and (4.26). We have $E_u(\sigma_x) = -E_g(\sigma_x)$ and the restriction of E_u to K_6 is nontrivial. The 6-dimensional space $\tilde{\mathsf{V}}_6$ is spanned by the tensors (4.24)–(4.26).

Finally, the 9-dimensional space $\tilde{\mathsf{V}}_6$ is spanned by the tensors (4.24)–(4.28).

Put $K = O \times Z_2^c$. Equation (4.10) gives $\tilde{\theta} = 2A_{1g} \oplus 2E_g$. The spaces $\tilde{\mathsf{V}}_0, \ldots, \tilde{\mathsf{V}}_7$ are intersections of the spaces of the previous case with the space of 3×3 symmetric matrices with real entries. That is, the tensors (4.24) constitute the basis of the spaces $\tilde{\mathsf{V}}_0$ and $\tilde{\mathsf{V}}_2$, the tensors (4.24), $\mathsf{f}^{E_g,12}$ of (4.25), and $\mathsf{f}^{E_g,22}$ of (4.26) are the basis of $\tilde{\mathsf{V}}_1$ and $\tilde{\mathsf{V}}_5$, the tensors (4.24) and (4.29) for $\tilde{\mathsf{V}}_3$ and $\tilde{\mathsf{V}}_4$, the tensors (4.24)–(4.26) for $\tilde{\mathsf{V}}_6$ and $\tilde{\mathsf{V}}_7$.

Finally, put $K = O^-$. Equation (4.16) gives $\tilde{\theta} = 2A_{1g} \oplus 2E_g \oplus A_{1u} \oplus E_u$. Again, all the spaces $\tilde{\mathsf{V}}_0, \ldots, \tilde{\mathsf{V}}_7$ contain the subspace $\tilde{\mathsf{V}}_0$ generated by the tensors (4.24).

We only need to analyse the irreducible component A_{1u}, because the rest of the components have already been analysed. By [2, Table 71.4], the restriction of A_{1u} to the groups K_1–K_6 is nontrivial, while that to K_7 is trivial. To construct the spaces $\tilde{\mathsf{V}}_1$–$\tilde{\mathsf{V}}_6$, we take the spaces from the case of $K = O$ with the same names, and delete the tensor (4.27), if it exists. The space $\tilde{\mathsf{V}}_1$ is spanned by the tensors (4.24), $\mathsf{f}^{E_g,12}$ of (4.25), and $\mathsf{f}^{E_g,22}$ of (4.26). $\tilde{\mathsf{V}}_2$ is spanned by (4.24), the spaces $\tilde{\mathsf{V}}_3$ and $\tilde{\mathsf{V}}_4$ by (4.24) and (4.29), the space $\tilde{\mathsf{V}}_5$ by (4.24), $\mathsf{f}^{E_g,12}$ of (4.25), $\mathsf{f}^{E_g,22}$ of (4.26), and $\mathsf{f}^{E_u,11}$ of (4.28). The space $\tilde{\mathsf{V}}_6$ is generated by (4.24)–(4.26), and $\tilde{\mathsf{V}}_7$ by (4.24)–(4.28).

4.6 Convex Compacta

Recall that the matrix $f(\vec{\lambda}_m)$ in (4.20) is Hermitian, nonnegative-definite and has unit trace. The set of such matrices is convex and compact. The intersection of the above set with the linear space $\tilde{\mathsf{V}}_m$ is a convex compact set, call it C_m. The function $f(\vec{\lambda}_m)$ is an arbitrary measurable function acting from $(\hat{\mathbb{R}}^3/\tilde{K})_m$ to C_m.

A point $f \in C_m$ is called *extreme* if it does not lie in any open line segment joining two points of C_m. It is well known that any convex compact set in a real finite-dimensional space is the closed convex hull of the set of its own extreme points. In the following example, we describe each set C_m and give a geometrical description of its set of extreme points, if possible.

Example 4.5 Let $G = Z_1$, $K = \mathrm{O}(3)$. Put $m = 0$. The set C_0 is the set of symmetric nonnegative-definite matrices with unit trace lying in the 5-dimensional space with basis (4.21). It turns out that C_0 is the simplex with five vertices. For details, see [20].

When $m = 1$, the set C_1 is the set of symmetric nonnegative-definite matrices with unit trace lying in some 21-dimensional space. The set of its extreme points has 3 connected components, see [20].

Until the end of Example 4.5 put $G = D_2$. If $K = D_2$ and $0 \leq m \leq 6$, then a point in C_m is a linear combination of the matrices (4.22), that is,

$$\mathsf{f} = \sum_{i=1}^{6} \nu_{A_g,i1} \mathsf{f}^{A_g,i1} = \begin{pmatrix} \nu_{A_g,11} & \frac{1}{\sqrt{2}}\nu_{A_g,61} & \frac{1}{\sqrt{2}}\nu_{A_g,51} \\ \frac{1}{\sqrt{2}}\nu_{A_g,61} & \nu_{A_g,21} & \frac{1}{\sqrt{2}}\nu_{A_g,41} \\ \frac{1}{\sqrt{2}}\nu_{A_g,51} & \frac{1}{\sqrt{2}}\nu_{A_g,41} & \nu_{A_g,31} \end{pmatrix}, \tag{4.30}$$

where the principal minors are nonnegative and the trace is equal to 1. The same convex compact set occurs when

- $K = D_2 \times Z_2^c$ for all m;
- $K = D_4$ and $m = 3$, $K = D_4^h$ and $m \in \{3, 6\}$, but $\nu_{A_g,k1}$ are replaced with $\nu_{A_{1g},k1}$ for $1 \leq k \leq 4$, and with $\nu_{B_{1g},k-4\,1}$ for $5 \leq k \leq 6$;
- $K = D_4 \times Z_2^c$ and $m \in \{6, 7\}$, but $\nu_{A_g,k1}$ are replaced with $\nu_{A_{1g},k1}$ for $1 \leq k \leq 4$, with $\nu_{A_{2g},11}$ for $k = 4$, with $\nu_{B_{2g},11}$ for $k = 5$, and with $\nu_{B_{1g},11}$ for $k = 6$.

The set C_7 is even larger. It consists of the matrices

$$\mathsf{f} = \begin{pmatrix} \nu_{A_g,11} & \frac{1}{\sqrt{2}}(\nu_{A_g,61} + \mathrm{i}\nu_{A_u,31}) & \frac{1}{\sqrt{2}}(\nu_{A_g,51} + \mathrm{i}\nu_{A_u,21}) \\ \frac{1}{\sqrt{2}}(\nu_{A_g,61} - \mathrm{i}\nu_{A_u,31}) & \nu_{A_g,21} & \frac{1}{\sqrt{2}}(\nu_{A_g,41} + \mathrm{i}\nu_{A_u,11}) \\ \frac{1}{\sqrt{2}}(\nu_{A_g,51} - \mathrm{i}\nu_{A_u,21}) & \frac{1}{\sqrt{2}}(\nu_{A_g,41} - \mathrm{i}\nu_{A_u,11}) & \nu_{A_g,31} \end{pmatrix} \tag{4.31}$$

with similar conditions. The same convex compact set occurs when

- $K = D_4$ and $m = 7$, but $\nu_{A_g,k1}$ are replaced with $\nu_{A_{1g},k1}$ for $1 \le k \le 4$, and with $\nu_{B_{1g},k-4\,1}$ for $5 \le k \le 6$, $\nu_{A_u,k1}$ is replaced with $\nu_{A_{1u},k1}$ for $k = 1$, and with $\nu_{B_{1u},k-1\,1}$ for $2 \le k \le 3$.
- $K = D_4^h$ and $m \in \{5, 7\}$, but $\nu_{A_g,k1}$ are replaced with $\nu_{A_{1g},k1}$ for $1 \le k \le 4$, and with $\nu_{B_{1g},k-4\,1}$ for $5 \le k \le 6$, $\nu_{A_u,k1}$ is replaced with $\nu_{A_{2u},k1}$ for $k = 1$, and with $\nu_{B_{2u},k-1\,1}$ for $2 \le k \le 3$.

When $K = D_4$ and $m \in \{0, 1, 2, 5, 6\}$, a point in C_m is a linear combination of the matrices (4.22a) and (4.22b), that is,

$$\mathsf{f} = \begin{pmatrix} \nu_{A_{1g},11} & 0 & 0 \\ 0 & \nu_{A_{1g},21} & \frac{1}{\sqrt{2}}\nu_{A_{1g},41} \\ 0 & \frac{1}{\sqrt{2}}\nu_{A_{1g},41} & \nu_{A_{1g},31} \end{pmatrix}. \tag{4.32}$$

The principal minors of this matrix are nonnegative and its trace is equal to 1 if and only if

$$\nu_{A_{1g},11} \ge 0, \qquad -\nu_{A_{1g},21} + \nu_{A_{1g},11}\nu_{A_{1g},21} + \nu^2_{A_{1g},21} + \frac{1}{2}\nu^2_{A_{1g},41} \le 0\,.$$

The set C_m is a closed cone. Its set of extreme points contains 2 components: the vertex with $\nu_{A_{1g},11} = 1$ and $\nu_{A_{1g},21} = \nu_{A_{1g},31} = \nu_{A_{1g},41} = 0$, and the circle with $\nu_{A_{1g},11} = 0$ and $(2\nu_{A_{1g},21} - 1)^2 + 2\nu^2_{A_{1g},41} = 1$. The same convex compact set occurs when $K = D_4^h$ and $m \in \{0, 2\}$.

When $K = D_4$ and $m = 4$, a point in C_4 is a linear combination of the matrices (4.22a), (4.22b), (4.23b), and (4.23c), that is

$$\mathsf{f} = \begin{pmatrix} \nu_{A_{1g},11} & \frac{\mathrm{i}}{\sqrt{2}}\nu_{A_{1u},31} & \frac{\mathrm{i}}{\sqrt{2}}\nu_{A_{1u},21} \\ -\frac{\mathrm{i}}{\sqrt{2}}\mathrm{i}\nu_{A_{1u},31} & \nu_{A_{1g},21} & \frac{1}{\sqrt{2}}\nu_{A_{1g},41} \\ -\frac{\mathrm{i}}{\sqrt{2}}\nu_{A_{1u},21} & \frac{1}{\sqrt{2}}\nu_{A_{1g},41} & \nu_{A_{1g},31} \end{pmatrix}. \tag{4.33}$$

This convex compact set is similar to the compact set (4.30), but its location in the 9-dimensional linear space of Hermitian 3×3 matrices is different.

When $K = D_4^h$ and $m \in \{1, 4\}$, a point in C_m is a linear combination of the matrices (4.22a), (4.22b), and (4.23a), that is

$$\mathsf{f} = \begin{pmatrix} \nu_{A_{1g},11} & 0 & 0 \\ 0 & \nu_{A_{1g},21} & \frac{1}{\sqrt{2}}(\nu_{A_{1g},41} + \mathrm{i}\nu_{A_{2u},11}) \\ 0 & \frac{1}{\sqrt{2}}(\nu_{A_{1g},41} - \mathrm{i}\nu_{A_{2u},11}) & \nu_{A_{1g},31} \end{pmatrix}. \tag{4.34}$$

The principal minors of this matrix are nonnegative and its trace is equal to 1 if and only if C_m is a cone. Its set of extreme points contains 2 components: the vertex with $\nu_{A_{1g},11} = 1$ and $\nu_{A_{1g},i2} = \nu_{A_{1g},41} = \nu_{A_{2u},11} = 0$, and the sphere with $\nu_{A_{1g},11} = 0$ and $(2\nu_{A_{1g},21} - 1)^2 + 2\nu^2_{A_{1g},41} + 2\nu^2_{A_{2u},11} = 1$.

When $K = D_4 \times Z_2^c$ and $m \in \{0, 1\}$, a point in C_m is a linear combination of the matrices (4.22a), that is,

$$\mathsf{f} = \begin{pmatrix} \nu_{A_{1g},11} & 0 & 0 \\ 0 & \nu_{A_{1g},21} & 0 \\ 0 & 0 & \nu_{A_{1g},31} \end{pmatrix}. \tag{4.35}$$

It is a triangle.

When $K = D_4 \times Z_2^c$ and $m \in \{2, 4\}$, a point in C_m is a linear combination of the matrices (4.22a) and (4.22c), that is,

$$\mathsf{f} = \begin{pmatrix} \nu_{A_{1g},11} & 0 & \frac{1}{\sqrt{2}}\nu_{B_{2g},11} \\ 0 & \nu_{A_{1g},21} & 0 \\ \frac{1}{\sqrt{2}}\nu_{B_{2g},11} & 0 & \nu_{A_{1g},31} \end{pmatrix}. \tag{4.36}$$

This convex compact set is similar to the compact set (4.32), but its location in the 9-dimensional linear space of Hermitian 3×3 matrices is different.

Another incarnation of the convex compact set (4.32) occurs when $K = D_4 \times Z_2^c$ and $m \in \{3, 5\}$. A point in C_m is a linear combination of the matrices (4.22a) and (4.22d), that is,

$$\mathsf{f} = \begin{pmatrix} \nu_{A_{1g},11} & \frac{1}{\sqrt{2}}\nu_{B_{1g},11} & 0 \\ \frac{1}{\sqrt{2}}\nu_{B_{1g},11} & \nu_{A_{1g},21} & 0 \\ 0 & 0 & \nu_{A_{1g},31} \end{pmatrix}. \tag{4.37}$$

When $K = \mathcal{T}$ and $m = 0$, a point in C_0 is a linear combination of the matrices (4.24), that is,

$$\mathsf{f} = \begin{pmatrix} \nu_{A_g,11} & 0 & 0 \\ 0 & \frac{1}{\sqrt{2}}\nu_{A_g,21} & 0 \\ 0 & 0 & \frac{1}{\sqrt{2}}\nu_{A_g,21} \end{pmatrix}. \tag{4.38}$$

It is a closed interval. The same set appears when

- $K = \mathcal{T} \times Z_2^c$ and $m \in \{0, 2\}$;
- $K = \mathcal{O}$ and $m = 0$, $K \in \{\mathcal{O}^-, \mathcal{O} \times Z_2^c\}$ and $m \in \{0, 2\}$, but this time $\nu_{A_g,i1}$ are replaced with $\nu_{A_{1g},i1}$.

When $K = \mathcal{T}$ and $m \in \{1, 3\}$, a point in C_0 is a linear combination of the matrices (4.24)–(4.26), that is

$$\mathsf{f}=\begin{pmatrix}\nu_{A_g,11} & \frac{1}{\sqrt{2}}\nu_{E_g,21} & \frac{1}{\sqrt{2}}\nu_{E_g,22}\\ \frac{1}{\sqrt{2}}\nu_{E_g,21} & \frac{1}{\sqrt{2}}(\nu_{A_g,21}-\nu_{E_g,11}) & \frac{1}{\sqrt{2}}\nu_{E_g,12}\\ \frac{1}{\sqrt{2}}\nu_{E_g,22} & \frac{1}{\sqrt{2}}\nu_{E_g,12} & \frac{1}{\sqrt{2}}(\nu_{A_g,21}+\nu_{E_g,11})\end{pmatrix}. \tag{4.39}$$

This is the same convex compact set as (4.30), but this time it is written in the coordinates (4.24)–(4.26). This set also appears when

- $K=\mathcal{T}\times Z_2^c$ and $m\in\{1,3,4,5\}$;
- $K\in\{\mathcal{O},\mathcal{O}^-,\mathcal{O}\times Z_2^c\}$ and $m=6$, $K=\mathcal{O}\times Z_2^c$ and $m=7$, but $\nu_{A_g,i1}$ is replaced with $\nu_{A_{1g},i1}$.

When $K=\mathcal{T}$ and $m=2$, a point in C_2 is a linear combination of the matrices (4.24) and (4.27), that is,

$$\mathsf{f}=\begin{pmatrix}\nu_{A_g,11} & 0 & 0\\ 0 & \frac{1}{\sqrt{2}}\nu_{A_g,21} & \frac{\mathrm{i}}{\sqrt{2}}\nu_{A_u,11}\\ 0 & \frac{\mathrm{i}}{\sqrt{2}}\nu_{A_u,11} & \frac{1}{\sqrt{2}}\nu_{A_g,21}\end{pmatrix}. \tag{4.40}$$

This is not so easy to see, but it is a triangle similar to (4.35) with vertices

$$\begin{pmatrix}1&0&0\\0&0&0\\0&0&0\end{pmatrix},\quad \frac{1}{2}\begin{pmatrix}0&0&0\\0&1&\mathrm{i}\\0&\mathrm{i}&1\end{pmatrix},\quad\text{and}\quad \frac{1}{2}\begin{pmatrix}0&0&0\\0&1&-\mathrm{i}\\0&-\mathrm{i}&1\end{pmatrix}.$$

The same set appears when $K\in\{\mathcal{O},\mathcal{O}\times Z_2^c\}$ and $m=2$.

When $K=\mathcal{T}$ and $m\in\{4,5\}$, a point in C_4 is a linear combination of the matrices (4.24)–(4.28), that is

$$\mathsf{f}=\begin{pmatrix}\nu_{A_g,11} & \frac{1}{\sqrt{2}}\nu_{E_g,21}+\frac{\mathrm{i}}{\sqrt{2}}\nu_{E_u,11} & \frac{1}{\sqrt{2}}\nu_{E_g,22}+\frac{\mathrm{i}}{\sqrt{2}}\nu_{E_u,12}\\ \frac{1}{\sqrt{2}}\nu_{E_g,21}-\frac{\mathrm{i}}{\sqrt{2}}\nu_{E_u,11} & \frac{1}{\sqrt{2}}(\nu_{A_g,21}-\nu_{E_g,11}) & \frac{1}{\sqrt{2}}\nu_{E_g,12}+\frac{\mathrm{i}}{\sqrt{2}}\nu_{A_u,11}\\ \frac{1}{\sqrt{2}}\nu_{E_g,22}-\frac{\mathrm{i}}{\sqrt{2}}\nu_{E_u,12} & \frac{1}{\sqrt{2}}\nu_{E_g,11}-\frac{\mathrm{i}}{\sqrt{2}}\nu_{A_u,11} & \frac{1}{\sqrt{2}}(\nu_{A_g,21}+\nu_{E_g,11})\end{pmatrix}. \tag{4.41}$$

This is the same convex compact set as (4.31), but this time it is written in the coordinates (4.24)–(4.28). This set also appears when $K\in\{\mathcal{O},\mathcal{O}^-\}$ and $m=7$, but $\nu_{A_g,i1}$ is replaced with $\nu_{A_{1g},i1}$ and $\nu_{A_u,i1}$ is replaced with $\nu_{A_{1u},11}$.

Another convex compact set appears when $K=\mathcal{O}$ and $m=1$. A point in C_1 is a linear combination of the matrices (4.24), $\mathsf{f}^{E_g,12}$ of (4.25) and $\mathsf{f}^{E_g,22}$ of (4.26), that is

$$\mathsf{f}=\begin{pmatrix}\nu_{A_g,11} & 0 & \frac{1}{\sqrt{2}}\nu_{E_g,22}\\ 0 & \frac{1}{\sqrt{2}}\nu_{A_g,21} & \frac{1}{\sqrt{2}}\nu_{E_g,12}\\ \frac{1}{\sqrt{2}}\nu_{E_g,22} & \frac{1}{\sqrt{2}}\nu_{E_g,12} & \frac{1}{\sqrt{2}}\nu_{A_g,21}\end{pmatrix}. \tag{4.42}$$

This set also appears when $K\in\{\mathcal{O}^-,\mathcal{O}\times Z_2^c\}$ and $m=1$, but $\nu_{A_g,i1}$ is replaced with $\nu_{A_{1g},i1}$.

When $K = O$ and $m = 3$, a point in C_3 is a linear combination of the matrices (4.24) and (4.29), that is,

$$\mathsf{f} = \begin{pmatrix} \nu_{A_g,11} & \frac{1}{2\sqrt{2}}\nu_{E_g,12} & -\frac{\sqrt{3}}{2\sqrt{2}}\nu_{E_g,12} \\ \frac{1}{2\sqrt{2}}\nu_{E_g,12} & \frac{1}{\sqrt{2}}\nu_{A_g,21} - \frac{1}{2\sqrt{2}}\nu_{E_g,11} & \frac{1}{\sqrt{2}}\nu_{E_g,12} \\ -\frac{\sqrt{3}}{2\sqrt{2}}\nu_{E_g,12} & \frac{1}{\sqrt{2}}\nu_{E_g,12} & \frac{1}{\sqrt{2}}\nu_{A_g,21} + \frac{1}{2\sqrt{2}}\nu_{E_g,11} \end{pmatrix}. \quad (4.43)$$

This set also appears when $K \in \{O^-, O \times Z_2^c\}$ and $m \in \{3, 4\}$, $K = O \times Z_2^c$ and $m = 5$, but $\nu_{A_g,i1}$ is replaced with $\nu_{A_{1g},i1}$.

The convex compact set from the case of $K = O$ and $m = 4$ is a linear combination of the matrices (4.24), (4.27), and (4.29), that is,

$$\mathsf{f} = \begin{pmatrix} \nu_{A_g,11} & \frac{1}{2\sqrt{2}}\nu_{E_g,12} & -\frac{\sqrt{3}}{2\sqrt{2}}\nu_{E_g,12} \\ \frac{1}{2\sqrt{2}}\nu_{E_g,12} & \frac{1}{\sqrt{2}}\nu_{A_g,21} - \frac{1}{2\sqrt{2}}\nu_{E_g,11} & \frac{1}{\sqrt{2}}\nu_{E_g,12} + \frac{\mathrm{i}}{\sqrt{2}}\nu_{A_u,11} \\ -\frac{\sqrt{3}}{2\sqrt{2}}\nu_{E_g,12} & \frac{1}{\sqrt{2}}\nu_{E_g,12} - \frac{\mathrm{i}}{\sqrt{2}}\nu_{A_u,11} & \frac{1}{\sqrt{2}}\nu_{A_g,11} + \frac{1}{2\sqrt{2}}\nu_{E_g,11} \end{pmatrix}, \quad (4.44)$$

while that from the case of $K = O$ and $m = 5$ is a linear combination of the matrices (4.24), (4.27), $\mathsf{f}^{E_u,11}$ of (4.28), and (4.29), that is,

$$\mathsf{f} = \begin{pmatrix} \nu_{A_g,11} & \frac{1}{2\sqrt{2}}\nu_{E_g,12} + \frac{\mathrm{i}}{\sqrt{2}}\nu_{E_u,11} & -\frac{\sqrt{3}}{2\sqrt{2}}\nu_{E_g,12} \\ \frac{1}{2\sqrt{2}}\nu_{E_g,12} - \frac{\mathrm{i}}{\sqrt{2}}\nu_{E_u,11} & \frac{1}{\sqrt{2}}\nu_{A_g,21} - \frac{1}{2\sqrt{2}}\nu_{E_g,11} & \frac{1}{\sqrt{2}}\nu_{E_g,12} + \frac{\mathrm{i}}{\sqrt{2}}\nu_{A_u,11} \\ -\frac{\sqrt{3}}{2\sqrt{2}}\nu_{E_g,12} & \frac{1}{\sqrt{2}}\nu_{E_g,12} - \frac{\mathrm{i}}{\sqrt{2}}\nu_{A_u,11} & \frac{1}{\sqrt{2}}\nu_{A_g,21} + \frac{1}{2\sqrt{2}}\nu_{E_g,11} \end{pmatrix}. \quad (4.45)$$

In the case of $K = O^-$ and $m = 5$ it is a linear combination of the matrices (4.24), $\mathsf{f}^{E_u,11}$ of (4.28), and (4.29), that is,

$$\mathsf{f} = \begin{pmatrix} \nu_{A_g,11} & \frac{1}{2\sqrt{2}}\nu_{E_g,12} + \frac{\mathrm{i}}{\sqrt{2}}\nu_{E_u,11} & -\frac{\sqrt{3}}{2\sqrt{2}}\nu_{E_g,12} \\ \frac{1}{2\sqrt{2}}\nu_{E_g,12} - \frac{\mathrm{i}}{\sqrt{2}}\nu_{E_u,11} & \frac{1}{\sqrt{2}}\nu_{A_g,21} - \frac{1}{2\sqrt{2}}\nu_{E_g,11} & \frac{1}{\sqrt{2}}\nu_{E_g,12} \\ -\frac{\sqrt{3}}{2\sqrt{2}}\nu_{E_g,12} & \frac{1}{\sqrt{2}}\nu_{E_g,12} & \frac{1}{\sqrt{2}}\nu_{A_g,21} + \frac{1}{2\sqrt{2}}\nu_{E_g,11} \end{pmatrix}. \quad (4.46)$$

We summarise our findings in Table 4.2.

4.7 The "Spherical Bessel Functions"

Rewrite Eq. (4.19) in the following form:

$$\langle \mathsf{d}(\vec{x}), \mathsf{d}(\vec{y})\rangle_{pq} = \sum_{m=0}^{M-1} \int_{(\hat{\mathbb{R}}^3/\tilde{K})_m} \int_{\tilde{K}} \mathrm{e}^{\mathrm{i}(g\vec{\lambda}_m, \vec{y}-\vec{x})} (\tilde{\theta}(g) f(\vec{\lambda}_m))_{pq} \, \mathrm{d}g \, \mathrm{d}\mu_m(\vec{\lambda}_m), \quad (4.47)$$

Table 4.2 The convex compact sets C_m

K	m							
	0	1	2	3	4	5	6	7
D_2	(4.30)	(4.30)	(4.30)	(4.30)	(4.30)	(4.30)	(4.30)	(4.31)
$D_2 \times Z_2^c$	(4.30)	(4.30)	(4.30)	(4.30)	(4.30)	(4.30)	(4.30)	(4.30)
D_4	(4.32)	(4.32)	(4.32)	(4.30)	(4.33)	(4.32)	(4.32)	(4.31)
D_4^h	(4.32)	(4.34)	(4.32)	(4.30)	(4.34)	(4.31)	(4.30)	(4.31)
$D_4 \times Z_2^c$	(4.35)	(4.35)	(4.36)	(4.37)	(4.36)	(4.37)	(4.30)	(4.30)
$\mathcal{T}$	(4.38)	(4.39)	(4.40)	(4.39)	(4.41)	(4.41)	–	–
$\mathcal{T} \times Z_2^c$	(4.38)	(4.39)	(4.38)	(4.39)	(4.39)	(4.39)	–	–
$\mathcal{O}$	(4.38)	(4.42)	(4.40)	(4.43)	(4.44)	(4.45)	(4.39)	(4.41)
$\mathcal{O}^-$	(4.38)	(4.42)	(4.38)	(4.43)	(4.43)	(4.46)	(4.39)	(4.41)
$\mathcal{O} \times Z_2^c$	(4.38)	(4.42)	(4.38)	(4.43)	(4.43)	(4.43)	(4.39)	(4.39)

where $\mathrm{d}g$ is the probabilistic invariant measure on $\tilde{K}$. In this section, we calculate the inner integral.

To perform this task, consider the function $\mathrm{e}^{\mathrm{i}(g\vec{\lambda}_m, \vec{y}-\vec{x})}$ as a function of $g \in \tilde{K}$ under fixed $\vec{\lambda}_m$, $\vec{x}$, and $\vec{y}$. This function is continuous. The *Fine Structure Theorem* [13] states the following. Let $\hat{K}$ be the set of equivalence classes of irreducible orthogonal representations of the group $\tilde{K}$. Choose a basis in each space on which a representation $\theta \in \hat{K}$ acts, and consider the corresponding matrix $\theta_{kl}(g)$. For each $\theta \in \hat{K}$ there exists a nonempty subset $I_\theta \subseteq \{1, 2, \dots, \dim \theta\}$ such that the functions

$$\{\sqrt{\dim \varrho}\theta_{kl}(g) \colon \theta \in \hat{K}, 1 \le k \le \dim \varrho, l \in I_\theta\}$$

constitute an orthonormal basis in the Hilbert space of square-integrable functions on $\tilde{K}$. Note that for irreducible *unitary* representations, we have $I_\theta = \{1, 2, \dots, \dim \theta\}$ for all θ. The corresponding result is known as the *Peter–Weyl Theorem*.

In particular, the *continuous* function $\mathrm{e}^{\mathrm{i}(g\vec{\lambda}_m, \vec{y}-\vec{x})}$ has the *uniformly convergent* Fourier expansion

$$\mathrm{e}^{\mathrm{i}(g\vec{\lambda}_m, \vec{y}-\vec{x})} = \sum_{\theta \in \hat{K}} \dim \theta \sum_{k=1}^{\dim \theta} \sum_{l \in I_\theta} j_{kl}^{\theta}(\vec{\lambda}_m, \vec{y} - \vec{x})\theta_{kl}(g), \tag{4.48}$$

where

$$j_{kl}^{\theta}(\vec{\lambda}_m, \vec{y} - \vec{x}) = \int_{\tilde{K}} \mathrm{e}^{\mathrm{i}(g\vec{\lambda}_m, \vec{y}-\vec{x})}\theta_{kl}(g)\, \mathrm{d}g. \tag{4.49}$$

How to use this result? The matrix $\mathsf{f}_{pq}(\vec{\lambda}_m)$ may be written in the form

$$
\mathsf{f}_{pq}(\vec{\lambda}_m) = \sum_{\theta \in \hat{K}} \sum_{i=1}^{n(\theta,\tilde{\theta})} \sum_{k=1}^{\dim \theta} v_{\theta,ik}(\vec{\lambda}_m)\mathsf{f}_{pq}^{\theta,ik},
$$

where $n(\theta, \tilde{\theta})$ is the number of copies of θ inside $\tilde{\theta}$. The representation $\tilde{\theta}$ acts on this matrix as

$$
(\tilde{\theta}(g) f(\vec{\lambda}_m))_{pq} = \sum_{\theta \in \hat{K}} \sum_{i=1}^{n(\theta,\tilde{\theta})} \sum_{k=1}^{\dim \theta} \sum_{r=1}^{\dim \theta} \theta_{kr}(g) v_{\theta,ir}(\vec{\lambda}_m)\mathsf{f}_{pq}^{\theta,ik}. \tag{4.50}
$$

Substitute (4.48) and (4.50) into (4.47) and use the Fine Structure Theorem. After simple algebraic calculations we obtain

$$
\begin{aligned}
\langle \mathsf{d}(\vec{x}), \mathsf{d}(\vec{y}) \rangle = \sum_{m=0}^{M-1} \int_{(\hat{\mathbb{R}}^3/\tilde{K})_m} \sum_{\theta \in \hat{K}} \sum_{k=1}^{\dim \theta} \sum_{l \in I_\theta} \sum_{i=1}^{n(\theta,\tilde{\theta})} j_{kl}^{\theta}(\vec{\lambda}_m, \vec{y} - \vec{x}) \\
\times v_{\theta,il}(\vec{\lambda}_m)\mathsf{f}^{\theta,ik}\, \mathrm{d}\mu_m(\vec{\lambda}_m).
\end{aligned} \tag{4.51}
$$

In the following example, we calculate the functions $j_{kl}^{\theta}(\vec{\lambda}_m, \vec{y} - \vec{x})$.

Example 4.6 Let $G = Z_1$, $K = \mathrm{O}(3)$. The Fourier expansion (4.48) becomes the *Rayleigh expansion*:

$$
\mathrm{e}^{\mathrm{i}(\vec{k},\vec{z})} = 4\pi \sum_{\ell=0}^{\infty} \mathrm{i}^\ell j_\ell(\lambda r) \sum_{m=-\ell}^{\ell} S_m^\ell(\vartheta_{\vec{k}}, \varphi_{\vec{k}}) S_m^\ell(\vartheta_{\vec{z}}, \varphi_{\vec{z}}),
$$

where j_ℓ is the spherical Bessel function, $(r, \vartheta_{\vec{z}}, \varphi_{\vec{z}})$ are the spherical coordinates in the space domain, $(\lambda, \vartheta_{\vec{k}}, \varphi_{\vec{k}})$ are the spherical coordinates in the wavenumber domain, S_m^ℓ are the real-valued spherical harmonics, and $\vec{z} = \vec{x} - \vec{y}$. This is the reason for the notation (4.49). We will call these functions the "spherical Bessel functions" in the remaining cases as well.

Until the end of this example put $G = D_2$. Let $K = D_2$. It follows from (4.11) that $n(A_g, \tilde{\theta}) = 6$, $n(A_u, \tilde{\theta}) = 3$, and $n(\theta, \tilde{\theta}) = 0$ for the remaining irreducible representations of the group $\tilde{K} = D_2 \times Z_2^c$. We need to calculate only the spherical Bessel functions $j^{A_g}(\vec{\lambda}_m, \vec{z})$ and $j^{A_u}(\vec{\lambda}_m, \vec{z})$.

Let $m = 7$. By definition of the spherical Bessel function, we have

$$
j^{A_g}(\vec{\lambda}_m, \vec{z}) = \frac{1}{8} \sum_{g \in D_2 \times Z_2^c} \mathrm{e}^{\mathrm{i}(g\vec{k},\vec{z})} A_g(g).
$$

With the help of [2, Table 31.4] we obtain

$$
j^{A_g}(\vec{k}, \vec{z}) = \cos(k_1 z_1)\cos(k_2 z_2)\cos(k_3 z_3). \tag{4.52}
$$

Similarly,

$$j^{A_u}(\vec{k},\vec{z}) = -\mathrm{i}\sin(k_1z_1)\sin(k_2z_2)\sin(k_3z_3)\,. \tag{4.53}$$

For the remaining values of m, the spherical Bessel functions are given by the same formula due to their continuity.

When $K = D_2 \times Z_2^c$, no new cases appear.

Let $K = D_4$. It follows from (4.12) that $n(A_{1g},\tilde{\theta}) = 4$, $n(B_{1g},\tilde{\theta}) = n(B_{1u},\tilde{\theta}) = 2$, $n(A_{1u},\tilde{\theta}) = 1$. This time, we use [2, Table 33.4], and obtain

$$\begin{aligned}
j^{A_{1g}}(\vec{k},\vec{z}) &= \frac{1}{2}\cos(k_3z_3)[\cos(k_1z_1)\cos(k_2z_2) + \cos(k_1z_2)\cos(k_2z_1)]\,,\\
j^{B_{1g}}(\vec{k},\vec{z}) &= \frac{1}{2}\cos(k_3z_3)[\cos(k_1z_1)\cos(k_2z_2) - \cos(k_1z_2)\cos(k_2z_1)]\,,\\
j^{A_{1u}}(\vec{k},\vec{z}) &= -\frac{\mathrm{i}}{2}\sin(k_3z_3)[\sin(k_1z_1)\sin(k_2z_2) - \sin(k_1z_2)\sin(k_2z_1)]\,,\\
j^{B_{1u}}(\vec{k},\vec{z}) &= -\frac{\mathrm{i}}{2}\sin(k_3z_3)[\sin(k_1z_1)\sin(k_2z_2) + \sin(k_1z_2)\sin(k_2z_1)]\,.
\end{aligned} \tag{4.54}$$

When $K = D_4^h$, two new cases appear.

$$\begin{aligned}
j^{A_{2u}}(\vec{k},\vec{z}) &= -\frac{\mathrm{i}}{2}\sin(k_3z_3)[\sin(k_1z_1)\sin(k_2z_2) + \sin(k_1z_2)\sin(k_2z_1)]\,,\\
j^{B_{2u}}(\vec{k},\vec{z}) &= -\frac{\mathrm{i}}{2}\sin(k_3z_3)[\sin(k_1z_1)\sin(k_2z_2) - \sin(k_1z_2)\sin(k_2z_1)]\,.
\end{aligned} \tag{4.55}$$

When $K = D_4 \times Z_2^c$, two new cases appear.

$$\begin{aligned}
j^{A_{2g}}(\vec{k},\vec{z}) &= \frac{1}{2}\cos(k_3z_3)[-\sin(k_1z_1)\sin(k_2z_2) + \sin(k_1z_2)\sin(k_2z_1)]\,,\\
j^{B_{2g}}(\vec{k},\vec{z}) &= \frac{1}{2}\cos(k_3z_3)[-\sin(k_1z_1)\sin(k_2z_2) - \sin(k_1z_2)\sin(k_2z_1)]\,.
\end{aligned} \tag{4.56}$$

Let $K = \mathcal{T}$. It follows from (4.13) that $n(A_g,\tilde{\theta}) = n(E_g,\tilde{\theta}) = 2$, $n(A_u,\tilde{\theta}) = n(E_u,\tilde{\theta}) = 1$. For the case of A_g and A_u, we use [2, Table 72.4] and obtain:

$$\begin{aligned}
j^{A_g}(\vec{k},\vec{z}) &= \frac{1}{3}[\cos(k_1z_1)\cos(k_2z_2)\cos(k_3z_3) + \cos(k_1z_2)\cos(k_2z_3)\cos(k_3z_1)\\
&\quad + \cos(k_1z_3)\cos(k_2z_1)\cos(k_3z_2)]\,,\\
j^{A_u}(\vec{k},\vec{z}) &= \frac{\mathrm{i}}{3}[\sin(k_1z_1)\cos(k_2z_2)\cos(k_3z_3) + \sin(k_1z_2)\cos(k_2z_3)\cos(k_3z_1)\\
&\quad + \sin(k_1z_3)\cos(k_2z_1)\cos(k_3z_2)]\,.
\end{aligned} \tag{4.57}$$

We calculate the matrix entries of the representation E_g using the methods described in Example 4.4. The result is as follows: the representation E_g maps the elements E,

$C_{2x}, C_{2y}, C_{2z}, i, \sigma_x, \sigma_y$ and σ_z of the normal subgroup $D_2 \times Z_2^c$ of the group $\mathcal{T} \times Z_2^c$ to the 2×2 identity matrix, the elements C_{31}^+, C_{32}^+, C_{33}^+, C_{34}^+, S_{61}^-, S_{62}^-, S_{63}^- and S_{64}^- to the matrix $\frac{1}{2}\begin{pmatrix} -1 & \sqrt{3} \\ -\sqrt{3} & -1 \end{pmatrix}$, and the elements C_{31}^-, C_{32}^-, C_{33}^-, C_{34}^-, S_{61}^+, S_{62}^+, S_{63}^+ and S_{64}^+ to the matrix $\frac{1}{2}\begin{pmatrix} -1 & -\sqrt{3} \\ \sqrt{3} & -1 \end{pmatrix}$. The matrix entries of the first column of the spherical Bessel function $j^{E_g}(\vec{k}, \vec{z})$ are

$$\begin{aligned} j_{11}^{E_g}(\vec{k}, \vec{z}) &= \frac{1}{3} \cos(k_1 z_1) \cos(k_2 z_2) \cos(k_3 z_3) - \frac{1}{6} \cos(k_1 z_2) \cos(k_2 z_3) \cos(k_3 z_1) \\ &\quad - \frac{1}{6} \cos(k_1 z_3) \cos(k_2 z_1) \cos(k_3 z_2), \\ j_{21}^{E_g}(\vec{k}, \vec{z}) &= -\frac{1}{2\sqrt{3}} [\cos(k_1 z_2) \cos(k_2 z_3) \cos(k_3 z_1) \\ &\quad - \cos(k_1 z_3) \cos(k_2 z_1) \cos(k_3 z_2)]. \end{aligned} \tag{4.58}$$

Later we will see that only the first column of this matrix-valued function is important.

The representation E_u is as follows:

$$E_u(g) = \begin{cases} E_g(g), & \text{if } g \in \mathcal{T}, \\ -E_g(g), & \text{otherwise}. \end{cases}$$

The spherical Bessel function $j^{E_u}(\vec{k}, \vec{z})$ is

$$\begin{aligned} j_{11}^{E_u}(\vec{k}, \vec{z}) &= \frac{\mathrm{i}}{3} \sin(k_1 z_1) \cos(k_2 z_2) \cos(k_3 z_3) - \frac{\mathrm{i}}{6} \sin(k_1 z_2) \cos(k_2 z_3) \cos(k_3 z_1) \\ &\quad - \frac{\mathrm{i}}{6} \sin(k_1 z_3) \cos(k_2 z_1) \cos(k_3 z_2), \\ j_{21}^{E_u}(\vec{k}, \vec{z}) &= -\frac{\mathrm{i}}{2\sqrt{3}} [\sin(k_1 z_2) \cos(k_2 z_3) \cos(k_3 z_1) \\ &\quad - \sin(k_1 z_3) \cos(k_2 z_1) \cos(k_3 z_2)]. \end{aligned} \tag{4.59}$$

When $K = \mathcal{T} \times Z_2^c$, no new cases appear.

Let $K = O$. It follows from (4.14) that $n(A_{1g}, \tilde{\theta}) = n(E_g, \tilde{\theta}) = 2$, $n(A_{2u}, \tilde{\theta}) = n(E_u, \tilde{\theta}) = 1$. Using [2, Table 71.4], we obtain

$$\begin{aligned} j^{A_{1g}}(\vec{k}, \vec{z}) &= \frac{1}{6} \{\cos(k_1 z_1)[\cos(k_2 z_2) \cos(k_3 z_3) + \cos(k_2 z_3) \cos(k_3 z_2)] \\ &\quad + \cos(k_2 z_1)[\cos(k_1 z_3) \cos(k_3 z_2) + \cos(k_1 z_2) \cos(k_3 z_3)] \\ &\quad + \cos(k_3 z_1)[\cos(k_1 z_2) \cos(k_2 z_3) + \cos(k_1 z_3) \cos(k_2 z_2)]\}, \end{aligned} \tag{4.60}$$

$$\begin{aligned} j^{A_{2u}}(\vec{k}, \vec{z}) &= -\frac{\mathrm{i}}{6} \{\sin(k_1 z_1)[\sin(k_2 z_2) \sin(k_3 z_3) + \sin(k_2 z_3) \sin(k_3 z_2)] \\ &\quad + \sin(k_2 z_1)[\sin(k_1 z_3) \sin(k_3 z_2) + \sin(k_1 z_2) \sin(k_3 z_3)] \\ &\quad + \sin(k_3 z_1)[\sin(k_1 z_2) \sin(k_2 z_3) + \sin(k_1 z_3) \sin(k_2 z_2)]\}. \end{aligned}$$

We calculate the matrix entries of the representations E_g and E_u using the methods described in Example 4.4. The result is as follows: the representation E_g maps the elements $E, i, C_{2x}, C_{2y}, C_{2z}, \sigma_x, \sigma_y$ and σ_z to the identity matrix, the elements C_{31}^+, C_{32}^+, C_{33}^+, C_{34}^+, S_{61}^-, S_{62}^-, S_{63}^- and S_{64}^- to the matrix $\frac{1}{2}\begin{pmatrix} -1 & \sqrt{3} \\ -\sqrt{3} & -1 \end{pmatrix}$, the elements C_{31}^-, C_{32}^-, C_{33}^-, C_{34}^-, S_{61}^+, S_{62}^+, S_{63}^+ and S_{64}^+ to the matrix $\frac{1}{2}\begin{pmatrix} -1 & -\sqrt{3} \\ \sqrt{3} & -1 \end{pmatrix}$, the elements C_{4x}^+, C_{4x}^-, S_{4x}^-, S_{4x}^+, C'_{2d}, C'_{2f}, σ_{d4} and σ_{d6} to the matrix $\frac{1}{2}\begin{pmatrix} 1 & -\sqrt{3} \\ -\sqrt{3} & -1 \end{pmatrix}$, the elements C_{4y}^+, C_{4y}^-, S_{4y}^-, S_{4y}^+, C'_{2c}, C'_{2e}, σ_{d3} and σ_{d5} to the matrix $\begin{pmatrix} 1 & 0 \\ 0 & -1 \end{pmatrix}$, and the elements C_{4z}^+, C_{4z}^-, S_{4z}^-, S_{4z}^+, C'_{2a}, C'_{2b}, σ_{d1} and σ_{d2} to the matrix $\frac{1}{2}\begin{pmatrix} 1 & \sqrt{3} \\ \sqrt{3} & -1 \end{pmatrix}$. The representation E_u maps the elements E, C_{2x}, C_{2y} and C_{2z} to the identity matrix, the elements i, σ_x, σ_y and σ_z to the matrix $\begin{pmatrix} -1 & 0 \\ 0 & -1 \end{pmatrix}$, the elements C_{31}^+, C_{32}^+, C_{33}^+ and C_{34}^+ to the matrix $\frac{1}{2}\begin{pmatrix} -1 & \sqrt{3} \\ -\sqrt{3} & -1 \end{pmatrix}$, the elements S_{61}^-, S_{62}^-, S_{63}^- and S_{64}^- to the matrix $\frac{1}{2}\begin{pmatrix} 1 & -\sqrt{3} \\ \sqrt{3} & 1 \end{pmatrix}$, the elements C_{31}^-, C_{32}^-, C_{33}^- and C_{34}^- to the matrix $\frac{1}{2}\begin{pmatrix} -1 & -\sqrt{3} \\ \sqrt{3} & -1 \end{pmatrix}$, the elements S_{61}^+, S_{62}^+, S_{63}^+ and S_{64}^+ to the matrix $\frac{1}{2}\begin{pmatrix} 1 & \sqrt{3} \\ -\sqrt{3} & 1 \end{pmatrix}$, the elements C_{4x}^+, C_{4x}^-, C'_{2d} and C'_{2f} to the matrix $\frac{1}{2}\begin{pmatrix} 1 & -\sqrt{3} \\ -\sqrt{3} & -1 \end{pmatrix}$, the elements S_{4x}^-, S_{4x}^+, σ_{d4} and σ_{d6} to the matrix $\frac{1}{2}\begin{pmatrix} -1 & \sqrt{3} \\ \sqrt{3} & 1 \end{pmatrix}$, the elements C_{4y}^+, C_{4y}^-, C'_{2c} and C'_{2e} to the matrix $\begin{pmatrix} -1 & 0 \\ 0 & 1 \end{pmatrix}$, the elements S_{4y}^-, S_{4y}^+, σ_{d3} and σ_{d5} to the matrix $\begin{pmatrix} 1 & 0 \\ 0 & -1 \end{pmatrix}$, the elements C_{4z}^+, C_{4z}^-, C'_{2a} and C'_{2b} to the matrix $\frac{1}{2}\begin{pmatrix} 1 & \sqrt{3} \\ \sqrt{3} & -1 \end{pmatrix}$, and the elements S_{4z}^-, S_{4z}^+, σ_{d1} and σ_{d2} to the matrix $\frac{1}{2}\begin{pmatrix} -1 & -\sqrt{3} \\ -\sqrt{3} & 1 \end{pmatrix}$.

The spherical Bessel functions have the form

$$\begin{aligned}
j_{11}^{E_g}(\vec{k}, \vec{z}) &= \frac{1}{12}\{\cos(k_3 z_1)[\cos(k_1 z_3)\cos(k_2 z_2) - \cos(k_1 z_2)\cos(k_2 z_3)] \\
&\quad + \cos(k_3 z_2)[\cos(k_1 z_1)\cos(k_2 z_3) - \cos(k_1 z_3)\cos(k_2 z_1)] \\
&\quad + 2\cos(k_3 z_3)[\cos(k_1 z_1)\cos(k_2 z_2) + \cos(k_1 z_2)\cos(k_2 z_1)]\}, \\
j_{21}^{E_g}(\vec{k}, \vec{z}) &= \frac{\sqrt{3}}{12}\{\cos(k_3 z_1)[\cos(k_1 z_2)\cos(k_2 z_3) + \cos(k_1 z_3)\cos(k_2 z_2)] \\
&\quad - \cos(k_3 z_2)[\cos(k_1 z_3)\cos(k_2 z_1) + \cos(k_1 z_1)\cos(k_2 z_3)]\}, \qquad (4.61) \\
j_{11}^{E_u}(\vec{k}, \vec{z}) &= \frac{\mathrm{i}}{12}\{\sin(k_3 z_1)[\sin(k_1 z_2)\sin(k_2 z_3) - \sin(k_1 z_3)\sin(k_2 z_2)] \\
&\quad + \sin(k_3 z_2)[\sin(k_1 z_3)\sin(k_2 z_1) - \sin(k_1 z_1)\sin(k_2 z_3)] \\
&\quad + 2\sin(k_3 z_3)[\sin(k_1 z_1)\sin(k_2 z_2) + \sin(k_1 z_2)\sin(k_2 z_1)]\}, \\
j_{21}^{E_u}(\vec{k}, \vec{z}) &= \frac{\mathrm{i}\sqrt{3}}{12}\{-\sin(k_3 z_1)[\sin(k_1 z_2)\sin(k_2 z_3) - \sin(k_1 z_3)\sin(k_2 z_2)] \\
&\quad + \sin(k_3 z_2)[\sin(k_1 z_3)\sin(k_2 z_1) + \sin(k_1 z_1)\sin(k_2 z_3)]\}.
\end{aligned}$$

Let $K = O \times Z_2^c$. No new cases appear. Let $K = O^-$. It follows from (4.16) that $n(A_{1g}, \tilde{\theta}) = n(E_g, \tilde{\theta}) = 2, n(A_{1u}, \tilde{\theta}) = n(E_u, \tilde{\theta}) = 1$. Using [2, Table 71.4], we obtain

$$
\begin{aligned}
j^{A_{1u}}(\vec{k}, \vec{z}) = \frac{\mathrm{i}}{6}\{&\sin(k_1 z_1)[\sin(k_2 z_2)\sin(k_3 z_3) - \sin(k_2 z_3)\sin(k_3 z_2)] \\
&+ \sin(k_1 z_2)[\sin(k_2 z_1)\sin(k_3 z_3) - \sin(k_2 z_3)\sin(k_3 z_1)] \\
&+ \sin(k_1 z_3)[\sin(k_2 z_2)\sin(k_3 z_1) - \sin(k_2 z_1)\sin(k_3 z_2)].
\end{aligned}
\tag{4.62}
$$

4.8 Correlation Structures

To calculate the two-point correlation tensor of a homogeneous and isotropic random field, we use Eq. (4.51).

Note that a measurable function $\mathsf{f}^{K,m}(\vec{k})$ defined on the set $(\hat{\mathbb{R}}^3/\tilde{K})_m$ and taking values in the set C_m can be uniquely written as the sum

$$
\mathsf{f}^{K,m}(\vec{k}) = \sum_{\theta \in \hat{K}\,:\, n(\theta,\tilde{\theta}) \neq 0} \mathsf{f}^{K,m,\theta}(\vec{k}),
$$

where the function $\mathsf{f}^{K,m,\theta}(\vec{k})$ is calculated as follows: write down the function $\mathsf{f}^{K,m}(\vec{k})$ in the matrix notation, and replace all the entries $v_{\theta',il}(\vec{k})$ with $\theta' \neq \theta$ by zeros. For example, $\mathsf{f}^{D_2,7}(\vec{k}) = \mathsf{f}^{D_2,7,A_g}(\vec{k}) + \mathsf{f}^{D_2,7,A_u}(\vec{k})$, where

$$
\mathsf{f}^{D_2,7,A_g}(\vec{k}) = \begin{pmatrix}
v_{A_g,11}(\vec{k}) & \frac{1}{\sqrt{2}}v_{A_g,61}(\vec{k}) & \frac{1}{\sqrt{2}}v_{A_g,51}(\vec{k}) \\
\frac{1}{\sqrt{2}}v_{A_g,61}(\vec{k}) & v_{A_g,21}(\vec{k}) & \frac{1}{\sqrt{2}}v_{A_g,41}(\vec{k}) \\
\frac{1}{\sqrt{2}}v_{A_g,51}(\vec{k}) & \frac{1}{\sqrt{2}}v_{A_g,41}(\vec{k}) & v_{A_g,31}(\vec{k})
\end{pmatrix},
$$

$$
\mathsf{f}^{D_2,7,A_u}(\vec{k}) = \begin{pmatrix}
0 & \frac{\mathrm{i}}{\sqrt{2}}v_{A_u,31}(\vec{k}) & \frac{\mathrm{i}}{\sqrt{2}}v_{A_u,21}(\vec{k}) \\
-\frac{\mathrm{i}}{\sqrt{2}}v_{A_u,31}(\vec{k}) & 0 & \frac{\mathrm{i}}{\sqrt{2}}v_{A_u,11}(\vec{k}) \\
-\frac{\mathrm{i}}{\sqrt{2}}v_{A_u,21}(\vec{k}) & \frac{\mathrm{i}}{\sqrt{2}}v_{A_u,11}(\vec{k}) & 0
\end{pmatrix},
$$

see Eq. (4.31), referred to in Table 4.2 at the intersection of the row labelled "D_2" and the column labelled "7".

Theorem 4.2 *Let $\mathsf{d}(\vec{x})$ be a homogeneous and $(D_2, 3A)$-isotropic random field. We have*

$$
\langle \mathsf{d}(\vec{x}) \rangle = C_1\mathsf{d}^1 + C_2\mathsf{d}^2 + C_3\mathsf{d}^3, \qquad C_i \in \mathbb{R},
$$

where d^i are the tensors of the basis (3.6). In the above basis, the two-point correlation tensor of the field has the form

$$\begin{aligned}\langle \mathsf{d}(\vec{x}), \mathsf{d}(\vec{y}) \rangle &= \sum_{m=0}^{7} \int_{(\hat{\mathbb{R}}^3/D_2 \times Z_2^c)_m} j^{A_g}(\vec{k}, \vec{y} - \vec{x}) \mathsf{f}^{D_2,m,A_g}(\vec{k}) \, \mathrm{d}\mu_m(\vec{k}) \\ &+ \int_{(\hat{\mathbb{R}}^3/D_2 \times Z_2^c)_7} j^{A_u}(\vec{k}, \vec{y} - \vec{x}) \mathsf{f}^{D_2,7,A_u}(\vec{k}) \, \mathrm{d}\mu_7(\vec{k}),\end{aligned} \tag{4.63}$$

where the spherical Bessel function $j^{A_g}(\vec{k}, \vec{y} - \vec{x})$ *(resp.* $j^{A_u}(\vec{k}, \vec{y} - \vec{x})$*) is given by Eq. (4.52) (resp. Eq. (4.53)). The field has the form*

$$\mathsf{d}(\vec{x}) = C_1 \mathsf{d}^1 + C_2 \mathsf{d}^2 + C_3 \mathsf{d}^3 + \sum_{i=1}^{3} \sum_{m=0}^{7} \sum_{n=1}^{8} \int_{(\hat{\mathbb{R}}^3/D_2 \times Z_2^c)_m} j_n(\vec{k}, \vec{x}) \, \mathrm{d}Z_{imn}(\vec{k}) \mathsf{d}^i, \tag{4.64}$$

where $j_n(\vec{k}, \vec{x})$ *are 8 different combinations of cosines and sines of* k_1x_1*,* k_2x_2*, and* k_3x_3*, and where* $Z_{imn}(\vec{k})$ *are centred real-valued random measures on* $(\hat{\mathbb{R}}^3/D_2 \times Z_2^c)_m$ *with the following nonzero correlations: when* $(m, m') \neq (7, 7)$*, then*

$$\mathsf{E}[Z_{imn}(A) Z_{lm'n'}(B)] = \delta_{mm'} \delta_{nn'} \int_{A \cap B} \mathsf{f}_{il}^{D_2,m,A_g}(\vec{k}) \, \mathrm{d}\mu_m(\vec{k}),$$

otherwise

$$\mathsf{E}[Z_{i7n}(A) Z_{l7n}(B)] = \int_{A \cap B} \mathsf{f}_{il}^{D_2,7,A_g}(\vec{k}) \, \mathrm{d}\mu_m(\vec{k}),$$

$$\begin{aligned}\mathsf{E}[Z_{i71}(A) Z_{l78}(B)] &= -\mathsf{E}[Z_{i78}(A) Z_{l71}(B)] = -\mathsf{E}[Z_{i72}(A) Z_{l77}(B)] \\ &= \mathsf{E}[Z_{i77}(A) Z_{l72}(B)] = -\mathsf{E}[Z_{i73}(A) Z_{l76}(B)] \\ &= \mathsf{E}[Z_{i76}(A) Z_{l73}(B)] = \mathsf{E}[Z_{i74}(A) Z_{l75}(B)] \\ &= -\mathsf{E}[Z_{i75}(A) Z_{l74}(B)] = -\frac{1}{\mathrm{i}} \int_{A \cap B} \mathsf{f}_{il}^{D_2,7,A_u}(\vec{k}) \, \mathrm{d}\mu_7(\vec{k}).\end{aligned}$$

Proof Equation (4.63) is a particular case of Eq. (4.51). We enumerate the functions $j_n(\vec{k}, \vec{x})$ in the lexicographic order, starting from

$$j_1(\vec{k}, \vec{x}) = \cos(k_1 z_1) \cos(k_2 z_2) \cos(k_3 z_3).$$

Calculating the two-point correlation tensor of the random field (4.64) and taking into account the given correlation structure, we arrive at (4.63).

In what follows, the random measures Z are always uncorrelated when the values of the index m are different.

Theorem 4.3 *Let* $\mathsf{d}(\vec{x})$ *be a homogeneous and* $(D_2 \times Z_2^c, 3A_u)$*-isotropic random field. We have* $\langle \mathsf{d}(\vec{x}) \rangle = \mathsf{0}$. *In the basis (3.6), the two-point correlation tensor of the field has the form*

$$\langle \mathsf{d}(\vec{x}), \mathsf{d}(\vec{y}) \rangle = \int_{\hat{\mathbb{R}}^3/D_2 \times Z_2^c} j^{A_g}(\vec{k}, \vec{y} - \vec{x}) \mathsf{f}^{D_2,0,A_g}(\vec{k}) \, \mathrm{d}\mu(\vec{k}) \,,$$

where the spherical Bessel function $j^{A_g}(\vec{k}, \vec{y} - \vec{x})$ *is given by Eq. (4.52). The field has the form*

$$\mathsf{d}(\vec{x}) = \sum_{i=1}^{3} \sum_{n=1}^{8} \int_{\hat{\mathbb{R}}^3/D_2 \times Z_2^c} j_n(\vec{k}, \vec{x}) \, \mathrm{d}Z_{in}(\vec{k}) \mathsf{d}^i \,,$$

where $j_n(\vec{k}, \vec{x})$ *are 8 different combinations of cosines and sines of* $k_1 x_1$, $k_2 x_2$, *and* $k_3 x_3$, *and where* $Z_{in}(\vec{k})$ *are centred real-valued random measures on* $\hat{\mathbb{R}}^3/D_2 \times Z_2^c$ *with the nonzero correlations:*

$$\mathsf{E}[Z_{in}(A) Z_{ln'}(B)] = \delta_{nn'} \int_{A \cap B} \mathsf{f}_{il}^{D_2,0,A_g}(\vec{k}) \, \mathrm{d}\mu(\vec{k}) \,.$$

The proof is left to the reader.

Theorem 4.4 *Let* $\mathsf{d}(\vec{x})$ *be a homogeneous and* $(D_4, A_1 \oplus 2B_1)$*-isotropic random field. We have*

$$\langle \mathsf{d}(\vec{x}) \rangle = C \mathsf{d}^1 \,,$$

where d^1 *is the first tensor of the basis (3.7). In the above basis, the two-point correlation tensor of the field has the form*

$$\begin{aligned}
\langle \mathsf{d}(\vec{x}), \mathsf{d}(\vec{y}) \rangle = & \sum_{m=0}^{7} \int_{(\hat{\mathbb{R}}^3/D_4 \times Z_2^c)_m} j^{A_{1g}}(\vec{k}, \vec{y} - \vec{x}) \mathsf{f}^{D_4,m,A_{1g}}(\vec{k}) \, \mathrm{d}\mu_m(\vec{k}) \\
& + \sum_{m \in \{3,4,7\}} \int_{(\hat{\mathbb{R}}^3/D_4 \times Z_2^c)_m} j^{A_{1u}}(\vec{k}, \vec{y} - \vec{x}) \mathsf{f}^{D_4,m,A_{1u}}(\vec{k}) \, \mathrm{d}\mu_m(\vec{k}) \\
& + \int_{(\hat{\mathbb{R}}^3/D_4 \times Z_2^c)_7} \Big[j^{B_{1g}}(\vec{k}, \vec{y} - \vec{x}) \mathsf{f}^{D_4,7,B_{1g}}(\vec{k}) \\
& + j^{B_{1u}}(\vec{k}, \vec{y} - \vec{x}) \mathsf{f}^{D_4,7,B_{1u}}(\vec{k}) \Big] \, \mathrm{d}\mu_7(\vec{k}) \,,
\end{aligned} \tag{4.65}$$

where the spherical Bessel functions $j^{A_{1g}}(\vec{k}, \vec{y} - \vec{x})$, ..., $j^{B_{1u}}(\vec{k}, \vec{y} - \vec{x})$ *are given by Eq. (4.54). The field has the form*

$$\mathsf{d}(\vec{x}) = C\mathsf{d}^1 + \frac{1}{\sqrt{2}} \sum_{i=1}^{3} \sum_{m=0}^{7} \sum_{n=1}^{16} \int_{(\hat{\mathbb{R}}^3/D_4 \times Z_2^c)_m} j_n(\vec{k}, \vec{x}) \, \mathrm{d}Z_{imn}(\vec{k}) \mathsf{d}^i ,$$

where $j_n(\vec{k}, \vec{x})$ *are 8 different combinations of cosines and sines of* k_1x_1, k_2x_2, *and* k_3x_3 *for* $1 \le n \le 8$, *and of cosines and sines of* k_1x_2, k_2x_1, *and* k_3x_3 *for* $9 \le n \le 16$, *and where* $Z_{imn}(\vec{k})$ *are centred real-valued random measures on* $(\hat{\mathbb{R}}^3/D_4 \times Z_2^c)_m$ *with the following nonzero correlations. For* $0 \le m \le 6$

$$\mathsf{E}[Z_{imn}(A) Z_{lmn}(B)] = \int_{A \cap B} \mathsf{f}_{il}^{D_4, m, A_{1g}}(\vec{k}) \, \mathrm{d}\mu_m(\vec{k}) .$$

For $m \in \{3, 4\}$ *and* $k \in \{0, 1\}$

$$\begin{aligned}
\mathsf{E}[Z_{im8k+1}(A) Z_{lm8k+8}(B)] &= -\mathsf{E}[Z_{im8k+8}(A) Z_{lm8k+1}(B)] \\
&= -\mathsf{E}[Z_{im8k+2}(A) Z_{lm8k+7}(B)] \\
&= \mathsf{E}[Z_{im8k+7}(A) Z_{lm8k+2}(B)] \\
&= -\mathsf{E}[Z_{im8k+3}(A) Z_{lm8k+6}(B)] \\
&= \mathsf{E}[Z_{im8k+6}(A) Z_{lm8k+3}(B)] \\
&= \mathsf{E}[Z_{im8k+4}(A) Z_{lm8k+5}(B)] \\
&= -\mathsf{E}[Z_{im8k+5}(A) Z_{lm8k+4}(B)] \\
&= -\frac{1}{\mathrm{i}} \int_{A \cap B} \mathsf{f}_{il}^{D_4, m, A_{1u}}(\vec{k}) \, \mathrm{d}\mu_m(\vec{k}) .
\end{aligned}$$

For $m = 7$ *and for* $k \in \{0, 1\}$

$$\mathsf{E}[Z_{i7n}(A) Z_{l7n}(B)] = \begin{cases} \int_{A \cap B} [\mathsf{f}_{il}^{D_4, 7, A_{1g}}(\vec{k}) + \mathsf{f}_{il}^{D_4, 7, B_{1g}}(\vec{k})] \, \mathrm{d}\mu_m(\vec{k}), & 1 \le n \le 8 , \\ \int_{A \cap B} [\mathsf{f}_{il}^{D_4, 7, A_{1g}}(\vec{k}) - \mathsf{f}_{il}^{D_4, 7, B_{1g}}(\vec{k})] \, \mathrm{d}\mu_m(\vec{k}), & 9 \le n \le 16 , \end{cases}$$

as well as

$$\begin{aligned}
\mathsf{E}[Z_{i7\,8k+1}(A)Z_{l7\,8k+8}(B)] &= -\mathsf{E}[Z_{i7\,8k+8}(A)Z_{l7\,8k+1}(B)] \\
&= -\mathsf{E}[Z_{i7\,8k+2}(A)Z_{l7\,8k+7}(B)] \\
&= \mathsf{E}[Z_{i7\,8k+7}(A)Z_{l7\,8k+2}(B)] \\
&= -\mathsf{E}[Z_{i7\,8k+3}(A)Z_{l7\,8k+6}(B)] \\
&= \mathsf{E}[Z_{i7\,8k+6}(A)Z_{l7\,8k+3}(B)] \\
&= \mathsf{E}[Z_{i7\,8k+4}(A)Z_{l7\,8k+5}(B)] \\
&= -\mathsf{E}[Z_{i7\,8k+5}(A)Z_{l7\,8k+4}(B)] \\
&= -\frac{1}{\mathrm{i}}\int\limits_{A\cap B}[\mathsf{f}_{il}^{D_2,7,A_{1u}}(\vec{k}) + (-1)^k \mathsf{f}_{il}^{D_4,7,B_{1u}}(\vec{k})]\,\mathrm{d}\mu_7(\vec{k})\,.
\end{aligned}$$

In the first line of Eq. (4.65), the terms that correspond to $m \in \{0, 1, 2, 5, 6\}$ may be further simplified. Indeed, in this case the cone C_m is a union of closed intervals, or one-dimensional simplexes. One extreme point of each interval, say f^1, is the vertex with $\nu_{A_{1g},11} = 1$ and $\nu_{A_{1g},21} = \nu_{A_{1g},41} = 0$, another one, say $\mathsf{f}^2(\vec{k})$, runs over the disk with $\nu_{A_{1g},11} = 0$ and $(2\nu_{A_{1g},21} - 1)^2 + 2\nu^2_{A_{1g},41} \leq 1$. A point $\mathsf{f}^{D_4,m,A_{1g}}(\vec{k}) \in C_m$ may be represented as

$$\mathsf{f}^{D_4,m,A_{1g}}(\vec{k}) = \lambda(\vec{k})\mathsf{f}^1 + (1 - \lambda(p))\mathsf{f}^2(\vec{k})\,.$$

Introduce the measures

$$\Phi_{m1}(A) = \int\limits_A \lambda(\vec{k})\,\mathrm{d}\mu_m(\vec{k}), \qquad \Phi_{m2}(A) = \mu_m(A) - \Phi_{m1}(A)\,.$$

Then we have

$$\begin{aligned}
&\int\limits_{(\hat{\mathbb{R}}^3/D_4\times Z_2^c)_m} j^{A_{1g}}(\vec{k}, \vec{y} - \vec{x})\mathsf{f}^{D_4,m,A_{1g}}(\vec{k})\,\mathrm{d}\mu_m(\vec{k}) \\
&\quad = \int\limits_{(\hat{\mathbb{R}}^3/D_4\times Z_2^c)_m} j^{A_{1g}}(\vec{k}, \vec{y} - \vec{x})\mathsf{f}^1\mathrm{d}\Phi_{m1}(\vec{k}) \\
&\quad + \int\limits_{(\hat{\mathbb{R}}^3/D_4\times Z_2^c)_m} j^{A_{1g}}(\vec{k}, \vec{y} - \vec{x})\mathsf{f}^2(\vec{k})\,\mathrm{d}\Phi_{m2}(\vec{k})\,.
\end{aligned}$$

The "arbitrary function" f^1 of the first integral on the right-hand side takes values in the singleton. Therefore it is a constant. The arbitrary function $\mathsf{f}^2(\vec{k})$ of the second integral takes values in the two-dimensional disk, not in the three-dimensional cone, as before.

Another way to simplify the correlation structure of a homogeneous and isotropic random field is as follows. Inscribe a simplex into the convex compact set C_m. Allow the function f to take values only in the above simplex rather than in all of C_m. The corresponding integral becomes the sum of as many integrals as vertices in the simplex. Each integral does not contain arbitrary functions. The resulting formula describes only a sufficient condition for its right-hand side to be the two-point correlation tensor of a homogeneous and isotropic random field. The greater the Lebesgue measure of the inscribed simplex in comparison with that of C_m, the closer the sufficient condition is to a necessary condition. See the examples in [20].

Theorem 4.5 *Let* $\mathsf{d}(\vec{x})$ *be a homogeneous and* $(D_4^h, 2A_1 \oplus B_1)$*-isotropic random field. We have*

$$\langle \mathsf{d}(\vec{x}) \rangle = C_1 \mathsf{d}^1 + C_2 \mathsf{d}^2 ,$$

where d^1 *and* d^2 *are the tensors of the basis (3.7). In the above basis, the two-point correlation tensor of the field has the form*

$$\begin{aligned}
\langle \mathsf{d}(\vec{x}), \mathsf{d}(\vec{y}) \rangle = & \sum_{m=0}^{7} \int_{(\hat{\mathbb{R}}^3/D_4 \times Z_2^c)_m} j^{A_{1g}}(\vec{k}, \vec{y} - \vec{x}) \mathsf{f}^{D_4^h, m, A_{1g}}(\vec{k}) \, \mathrm{d}\mu_m(\vec{k}) \\
& + \sum_{m \in \{1,4,5,7\}} \int_{(\hat{\mathbb{R}}^3/D_4 \times Z_2^c)_m} \left[j^{A_{2u}}(\vec{k}, \vec{y} - \vec{x}) \mathsf{f}^{D_4^h, m, A_{2u}}(\vec{k}) \right] \mathrm{d}\mu_m(\vec{k}) \\
& + \sum_{m \in \{3,5,6,7\}} \int_{(\hat{\mathbb{R}}^3/D_4 \times Z_2^c)_m} \left[j^{B_{1g}}(\vec{k}, \vec{y} - \vec{x}) \mathsf{f}^{D_4^h, m, B_{1g}}(\vec{k}) \right] \mathrm{d}\mu_m(\vec{k}) \\
& + \sum_{m \in \{5,7\}} \int_{(\hat{\mathbb{R}}^3/D_4 \times Z_2^c)_m} \left[j^{B_{2u}}(\vec{k}, \vec{y} - \vec{x}) \mathsf{f}^{D_4^h, m, B_{2u}}(\vec{k}) \right] \mathrm{d}\mu_m(\vec{k}) ,
\end{aligned}$$

where the spherical Bessel functions $j^{A_{1g}}(\vec{k}, \vec{y} - \vec{x})$ *and* $j^{B_{1g}}(\vec{k}, \vec{y} - \vec{x})$ *are given by Eq. (4.54), and* $j^{A_{2u}}(\vec{k}, \vec{y} - \vec{x})$ *and* $j^{B_{2u}}(\vec{k}, \vec{y} - \vec{x})$ *are given by Eq. (4.55). The field has the form*

$$\mathsf{d}(\vec{x}) = C_1 \mathsf{d}^1 + C_2 \mathsf{d}^2 + \frac{1}{\sqrt{2}} \sum_{i=1}^{3} \sum_{m=0}^{7} \sum_{n=1}^{16} \int_{(\hat{\mathbb{R}}^3/D_4 \times Z_2^c)_m} j_n(\vec{k}, \vec{x}) \, \mathrm{d}Z_{imn}(\vec{k}) \mathsf{d}^i ,$$

where $j_n(\vec{k}, \vec{x})$ *are 8 different combinations of cosines and sines of* $k_1 x_1$*,* $k_2 x_2$*, and* $k_3 x_3$ *for* $1 \leq n \leq 8$*, and of cosines and sines of* $k_1 x_2$*,* $k_2 x_1$*, and* $k_3 x_3$ *for* $9 \leq n \leq 16$*, and where* $Z_{imn}(\vec{k})$ *are centred real-valued random measures on* $(\hat{\mathbb{R}}^3/D_4 \times Z_2^c)_m$ *with the following nonzero correlations. For* $m \in \{0, 1, 2, 4\}$

$$\mathsf{E}[Z_{imn}(A)Z_{lmn}(B)] = \int\limits_{A\cap B} \mathsf{f}_{il}^{D_4^h,m,A_{1g}}(\vec{k})\,\mathrm{d}\mu_m(\vec{k})\,.$$

For $m \in \{3,5,6,7\}$

$$\mathsf{E}[Z_{imn}(A)Z_{lmn}(B)] = \begin{cases} \int\limits_{A\cap B} [\mathsf{f}_{il}^{D_4^h,m,A_{1g}}(\vec{k}) + \mathsf{f}_{il}^{D_4,m,B_{1g}}(\vec{k})]\,\mathrm{d}\mu_m(\vec{k}), & 1 \le n \le 8,\\ \int\limits_{A\cap B} [\mathsf{f}_{il}^{D_4^h,m,A_{1g}}(\vec{k}) - \mathsf{f}_{il}^{D_4,m,B_{1g}}(\vec{k})]\,\mathrm{d}\mu_m(\vec{k}), & \textit{otherwise}. \end{cases}$$

For $m \in \{1,4\}$ *and* $k \in \{0,1\}$

$$\begin{aligned} \mathsf{E}[Z_{im8k+1}(A)Z_{lm8k+8}(B)] &= -\,\mathsf{E}[Z_{im8k+8}(A)Z_{lm8k+1}(B)]\\ &= -\mathsf{E}[Z_{im8k+2}(A)Z_{lm8k+7}(B)]\\ &= \mathsf{E}[Z_{im8k+7}(A)Z_{lm8k+2}(B)]\\ &= -\mathsf{E}[Z_{im8k+3}(A)Z_{lm8k+6}(B)]\\ &= \mathsf{E}[Z_{im8k+6}(A)Z_{lm8k+3}(B)]\\ &= \mathsf{E}[Z_{im8k+4}(A)Z_{lm8k+5}(B)]\\ &= -\mathsf{E}[Z_{im8k+5}(A)Z_{lm8k+4}(B)]\\ &= -\frac{1}{\mathrm{i}}\int\limits_{A\cap B} \mathsf{f}_{il}^{D_4^h,m,A_{2u}}(\vec{k})\,\mathrm{d}\mu_m(\vec{k})\,. \end{aligned}$$

For $m \in \{5,7\}$ *and* $k \in \{0,1\}$

$$\begin{aligned} \mathsf{E}[Z_{im8k+1}(A)Z_{lm8k+8}(B)] &= -\,\mathsf{E}[Z_{im8k+8}(A)Z_{lm8k+1}(B)]\\ &= -\mathsf{E}[Z_{im8k+2}(A)Z_{lm8k+7}(B)]\\ &= \mathsf{E}[Z_{im8k+7}(A)Z_{lm8k+2}(B)]\\ &= -\mathsf{E}[Z_{im8k+3}(A)Z_{lm8k+6}(B)]\\ &= \mathsf{E}[Z_{im8k+6}(A)Z_{lm8k+3}(B)]\\ &= \mathsf{E}[Z_{im8k+4}(A)Z_{lm8k+5}(B)]\\ &= -\mathsf{E}[Z_{im8k+5}(A)Z_{lm8k+4}(B)]\\ &= -\frac{1}{\mathrm{i}}\int\limits_{A\cap B} [\mathsf{f}_{il}^{D_4^h,m,A_{2u}}(\vec{k})\\ &\quad + (-1)^k \mathsf{f}_{il}^{D_4^h,m,B_{2u}}(\vec{k})]\,\mathrm{d}\mu_m(\vec{k})\,. \end{aligned}$$

Theorem 4.6 *Let* $\mathsf{d}(\vec{x})$ *be a homogeneous and* $(D_4 \times Z_2^c, A_{1u} \oplus B_{1u} \oplus B_{2u})$ *-isotropic random field. We have*

$$\langle \mathsf{d}(\vec{x})\rangle = \mathsf{0}\,.$$

In the basis (3.7), the two-point correlation tensor of the field has the form

$$
\begin{aligned}
\langle \mathsf{d}(\vec{x}), \mathsf{d}(\vec{y}) \rangle = \sum_{m=0}^{7} \int\limits_{(\hat{\mathbb{R}}^3/D_4\times Z_2^c)_m} & j^{A_{1g}}(\vec{k}, \vec{y}-\vec{x}) \mathsf{f}^{D_4\times Z_2^c, m, A_{1g}}(\vec{k}) \, \mathrm{d}\mu_m(\vec{k}) \\
+ \sum_{m\in\{6,7\}} \int\limits_{(\hat{\mathbb{R}}^3/D_4\times Z_2^c)_m} & j^{A_{2g}}(\vec{k}, \vec{y}-\vec{x}) \mathsf{f}^{D_4\times Z_2^c, m, B_{1g}}(\vec{k}) \\
+ \sum_{m\in\{3,5,6,7\}} \int\limits_{(\hat{\mathbb{R}}^3/D_4\times Z_2^c)_m} & j^{B_{1g}}(\vec{k}, \vec{y}-\vec{x}) \mathsf{f}^{D_4\times Z_2^c, m, B_{1g}}(\vec{k}) \, \mathrm{d}\mu_m(\vec{k}) \\
+ \sum_{m\in\{2,4,6,7\}} \int\limits_{(\hat{\mathbb{R}}^3/D_4\times Z_2^c)_m} & j^{B_{2g}}(\vec{k}, \vec{y}-\vec{x}) \mathsf{f}^{D_4\times Z_2^c, m, B_{2g}}(\vec{k}) \, \mathrm{d}\mu_m(\vec{k}) \,,
\end{aligned}
$$

where the spherical Bessel functions $j^{A_{1g}}(\vec{k}, \vec{y}-\vec{x})$ *and* $j^{B_{1g}}(\vec{k}, \vec{y}-\vec{x})$ *are given by Eq. (4.54),* $j^{A_{2g}}(\vec{k}, \vec{y}-\vec{x})$ *and* $j^{B_{2g}}(\vec{k}, \vec{y}-\vec{x})$ *are given by Eq. (4.56). The field has the form*

$$
\mathsf{d}(\vec{x}) = \frac{1}{\sqrt{2}} \sum_{i=1}^{3} \sum_{m=0}^{7} \sum_{n=1}^{16} \int\limits_{(\hat{\mathbb{R}}^3/D_4\times Z_2^c)_m} j_n(\vec{k}, \vec{x}) \, \mathrm{d}Z_{imn}(\vec{k}) \mathsf{d}^i \,,
$$

where $j_n(\vec{k}, \vec{x})$ *are 8 different combinations of cosines and sines of* k_1x_1, k_2x_2, *and* k_3x_3 *for* $1 \le n \le 8$, *and of cosines and sines of* k_1x_2, k_2x_1, *and* k_3x_3 *for* $9 \le n \le 16$, *and where* $Z_{imn}(\vec{k})$ *are centred real-valued random measures on* $(\hat{\mathbb{R}}^3/D_4 \times Z_2^c)_m$ *with the following nonzero correlations. For* $m \in \{0, 1, 2, 4\}$

$$
\mathsf{E}[Z_{imn}(A)Z_{lmn}(B)] = \int\limits_{A\cap B} \mathsf{f}_{il}^{D_4\times Z_2^c, m, A_{1g}}(\vec{k}) \, \mathrm{d}\mu_m(\vec{k}) \,.
$$

For $m \in \{3, 5, 6, 7\}$

$$
\begin{aligned}
&\mathsf{E}[Z_{imn}(A)Z_{lmn}(B)] \\
&= \begin{cases} \int_{A\cap B} [\mathsf{f}_{il}^{D_4\times Z_2^c, m, A_{1g}}(\vec{k}) + \mathsf{f}_{il}^{D_4\times Z_2^c, m, B_{1g}}(\vec{k})] \, \mathrm{d}\mu_m(\vec{k}), & 1 \le n \le 8 \,, \\ \int_{A\cap B} [\mathsf{f}_{il}^{D_4\times Z_2^c, m, A_{1g}}(\vec{k}) - \mathsf{f}_{il}^{D_4\times Z_2^c, m, B_{1g}}(\vec{k})] \, \mathrm{d}\mu_m(\vec{k}), & 9 \le n \le 16 \,. \end{cases}
\end{aligned}
$$

For $m \in \{2, 4\}$ *and* $k \in \{0, 1\}$

$$\begin{aligned}
-\mathsf{E}[Z_{im8k+1}(A)Z_{lm8k+7}(B)] &= -\mathsf{E}[Z_{im8k+7}(A)Z_{lm8k+1}(B)] \\
&= -\mathsf{E}[Z_{im8k+2}(A)Z_{lm8k+8}(B)] \\
&= -\mathsf{E}[Z_{im8k+8}(A)Z_{lm8k+2}(B)] \\
&= \mathsf{E}[Z_{im8k+3}(A)Z_{lm8k+5}(B)] \\
&= \mathsf{E}[Z_{im8k+5}(A)Z_{lm8k+3}(B)] \\
&= \mathsf{E}[Z_{im8k+4}(A)Z_{lm8k+6}(B)] \\
&= \mathsf{E}[Z_{im8k+6}(A)Z_{lm8k+4}(B)] \\
&= \int_{A\cap B} \mathsf{f}_{il}^{D_4\times Z_2^c,m,B_{2g}}(\vec{k})\,\mathrm{d}\mu_m(\vec{k})\,.
\end{aligned}$$

For $m \in \{6, 7\}$ *and* $k \in \{0, 1\}$

$$\begin{aligned}
-\mathsf{E}[Z_{im8k+1}(A)Z_{lm8k+7}(B)] &= -\mathsf{E}[Z_{im8k+7}(A)Z_{lm8k+1}(B)] \\
&= -\mathsf{E}[Z_{im8k+2}(A)Z_{lm8k+8}(B)] \\
&= -\mathsf{E}[Z_{im8k+8}(A)Z_{lm8k+2}(B)] \\
&= \mathsf{E}[Z_{im8k+3}(A)Z_{lm8k+5}(B)] \\
&= \mathsf{E}[Z_{im8k+5}(A)Z_{lm8k+3}(B)] \\
&= \mathsf{E}[Z_{im8k+4}(A)Z_{lm8k+6}(B)] \\
&= \mathsf{E}[Z_{im8k+6}(A)Z_{lm8k+4}(B)] \\
&= \int_{A\cap B} \Big[\mathsf{f}_{il}^{D_4\times Z_2^c,m,A_{2g}}(\vec{k}) \\
&\quad + (-1)^k \mathsf{f}_{il}^{D_4\times Z_2^c,m,B_{2g}}(\vec{k})\Big]\,\mathrm{d}\mu_m(\vec{k})\,.
\end{aligned}$$

Theorem 4.7 *Let* $\mathsf{d}(\vec{x})$ *be a homogeneous and* $(\mathcal{T}, A \oplus E)$*-isotropic random field. We have*

$$\langle \mathsf{d}(\vec{x})\rangle = C_1 \mathsf{d}^1\,,$$

where d^1 *is the first tensor of the basis (3.8). In the above basis, the two-point correlation tensor of the field has the form*

$$\begin{aligned}\langle \mathsf{d}(\vec{x}), \mathsf{d}(\vec{y})\rangle &= \sum_{m=0}^{5} \int\limits_{(\hat{\mathbb{R}}^3/\mathcal{T}\times Z_2^c)_m} j^{A_g}(\vec{k}, \vec{y}-\vec{x})\mathsf{f}^{\mathcal{T},m,A_g}(\vec{k})\,\mathrm{d}\mu_m(\vec{k})\\
&+ \int\limits_{(\hat{\mathbb{R}}^3/\mathcal{T}\times Z_2^c)_2} j^{A_u}(\vec{k}, \vec{y}-\vec{x})\mathsf{f}^{\mathcal{T},2,A_u}(\vec{k})\,\mathrm{d}\mu_2(\vec{k})\\
&+ \sum_{m\in\{1,3,4,5\}} \int\limits_{(\hat{\mathbb{R}}^3/\mathcal{T}\times Z_2^c)_m} \sum_{k=1}^{2} j^{E_g}_{k1}(\vec{k}, \vec{y}-\vec{x}) \sum_{q=1}^{2} \nu_{E_g,q1}(\vec{k})\mathsf{f}^{E_g,qk}\,\mathrm{d}\mu_m(\vec{k})\\
&+ \sum_{m=4}^{5} \int\limits_{(\hat{\mathbb{R}}^3/\mathcal{T}\times Z_2^c)_m} \nu_{E_u,11}(\vec{k}) \sum_{k=1}^{2} j^{E_u}_{k1}(\vec{k}, \vec{y}-\vec{x})\mathsf{f}^{E_u,1k}\,\mathrm{d}\mu_m(\vec{k})\,,\end{aligned}$$

where the spherical Bessel functions $j^{A_g}(\vec{k}, \vec{y}-\vec{x})$, ..., $j^{E_u}(\vec{k}, \vec{y}-\vec{x})$ *are given by Eqs. (4.57)–(4.59), and the basis tensors* $\mathsf{f}^{E_g,ik}$ *and* $\mathsf{f}^{E_u,1k}$ *are given by Eqs. (4.25), (4.26), and (4.28). The field has the form*

$$\mathsf{d}(\vec{x}) = C_1\mathsf{d}^1 + \frac{1}{\sqrt{3}}\sum_{i=1}^{3}\sum_{m=0}^{5}\sum_{n=1}^{24} \int\limits_{(\hat{\mathbb{R}}^3/\mathcal{T}\times Z_2^c)_m} j_n(\vec{k}, \vec{x})\,\mathrm{d}Z_{imn}(\vec{k})\mathsf{f}^i\,,$$

where $j_n(\vec{k}, \vec{x})$ *are 8 different combinations of cosines and sines of* k_1x_1, k_2x_2, *and* k_3x_3 *for* $1 \le n \le 8$, *of cosines and sines of* k_1x_2, k_2x_3, *and* k_3x_1 *for* $9 \le n \le 16$, *and of cosines and sines of* k_1x_3, k_2x_1, *and* k_3x_2 *for* $17 \le n \le 24$. $Z_{mn}(\vec{k})$ *are centred real-valued random measures on* $(\hat{\mathbb{R}}^3/\mathcal{T}\times Z_2^c)_m$ *with the following nonzero correlations. For* $m \in \{0, 2\}$

$$\mathsf{E}[Z_{imn}(A)Z_{lmn}(B)] = \int\limits_{A\cap B} \mathsf{f}^{\mathcal{T},m,A_g}_{il}(\vec{k})\,\mathrm{d}\mu_m(\vec{k})\,.$$

For $m \in \{1, 3, 4, 5\}$ *and* $1 \le n \le 8$

$$\mathsf{E}[Z_{imn}(A)Z_{lmn}(B)] = \int\limits_{A\cap B} \left[\mathsf{f}^{\mathcal{T},m,A_g}_{il}(\vec{k}) + \sum_{q=1}^{2} \nu_{E_g,q1}(\vec{k})\mathsf{f}^{E_g,q1}_{il}\right]\mathrm{d}\mu_m(\vec{k})\,.$$

For $m \in \{1, 3, 4, 5\}$ *and* $9 \leq n \leq 16$

$$\mathsf{E}[Z_{imn}(A)Z_{lmn}(B)] = \int_{A\cap B} \left[\mathsf{f}^{\mathcal{T},m,A_g}(\vec{k}) + \sum_{q=1}^{2} \nu_{E_g,q1}(\vec{k}) \left(-\frac{1}{2}\mathsf{f}_{il}^{E_g,q1} - \frac{\sqrt{3}}{2}\mathsf{f}_{il}^{E_g,q2}\right)\right] \mathrm{d}\mu_m(\vec{k})\,.$$

For $m \in \{1, 3, 4, 5\}$ *and* $17 \leq n \leq 24$

$$\mathsf{E}[Z_{imn}(A)Z_{lmn}(B)] = \int_{A\cap B} \left[\mathsf{f}^{\mathcal{T},m,A_g}(\vec{k}) + \sum_{q=1}^{2} \nu_{E_g,q1}(\vec{k}) \left(-\frac{1}{2}\mathsf{f}_{il}^{E_g,q1} + \frac{\sqrt{3}}{2}\mathsf{f}_{il}^{E_g,q2}\right)\right] \mathrm{d}\mu_m(\vec{k})\,.$$

For $m = 2$ *and* $0 \leq k \leq 2$

$$\begin{aligned}
\mathsf{E}[Z_{i2\,8k+1}(A)Z_{l2\,8k+5}(B)] &= -\mathsf{E}[Z_{i2\,8k+5}(A)Z_{l2\,8k+1}(B)] \\
&= \mathsf{E}[Z_{i2\,8k+2}(A)Z_{l2\,8k+6}(B)] \\
&= -\mathsf{E}[Z_{i2\,8k+6}(A)Z_{l2\,8k+2}(B)] \\
&= \mathsf{E}[Z_{i2\,8k+3}(A)Z_{l2\,8k+7}(B)] \\
&= -\mathsf{E}[Z_{i2\,8k+7}(A)Z_{l2\,8k+3}(B)] \\
&= \mathsf{E}[Z_{i2\,8k+4}(A)Z_{l2\,8k+8}(B)] \\
&= -\mathsf{E}[Z_{i2\,8k+8}(A)Z_{l2\,8k+4}(B)] \\
&= \int_{A\cap B} \mathsf{f}_{il}^{\mathcal{T},2,A_u}(\vec{k})\,\mathrm{d}\mu_2(\vec{k})\,.
\end{aligned}$$

For $m \in \{4, 5\}$ *and* $1 \leq n \leq 8$

$$\begin{aligned}
\mathsf{E}[Z_{im1}(A)Z_{lm5}(B)] &= -\mathsf{E}[Z_{im5}(A)Z_{lm1}(B)] = \mathsf{E}[Z_{im2}(A)Z_{lm6}(B)] \\
&= -\mathsf{E}[Z_{im6}(A)Z_{lm2}(B)] = \mathsf{E}[Z_{im3}(A)Z_{lm7}(B)] \\
&= -\mathsf{E}[Z_{im7}(A)Z_{lm3}(B)] = \mathsf{E}[Z_{im4}(A)Z_{lm8}(B)] \\
&= -\mathsf{E}[Z_{im8}(A)Z_{lm4}(B)] = \int_{A\cap B} \nu_{E_u,11}(\vec{k})\mathsf{f}_{il}^{E_u,11}\,\mathrm{d}\mu_m(\vec{k})\,.
\end{aligned}$$

For $m \in \{4, 5\}$ *and* $9 \le n \le 16$

$$\begin{aligned}
\mathsf{E}[Z_{im9}(A)Z_{lm13}(B)] &= -\mathsf{E}[Z_{im13}(A)Z_{lm9}(B)]\\
&= \mathsf{E}[Z_{im10}(A)Z_{lm14}(B)]\\
&= -\mathsf{E}[Z_{im14}(A)Z_{lm10}(B)]\\
&= \mathsf{E}[Z_{im11}(A)Z_{lm15}(B)]\\
&= -\mathsf{E}[Z_{im15}(A)Z_{lm11}(B)]\\
&= \mathsf{E}[Z_{im12}(A)Z_{lm16}(B)]\\
&= -\mathsf{E}[Z_{im16}(A)Z_{lm12}(B)]\\
&= \int\limits_{A\cap B} \nu_{E_u,11}(\vec{k}) \left(-\frac{1}{2}\mathsf{f}^{E_u,11}_{il} - \frac{\sqrt{3}}{2}\mathsf{f}^{E_u,12}_{il}\right) \mathrm{d}\mu_m(\vec{k})\,.
\end{aligned}$$

For $m \in \{4, 5\}$ *and* $17 \le n \le 24$

$$\begin{aligned}
\mathsf{E}[Z_{im17}(A)Z_{lm21}(B)] &= -\mathsf{E}[Z_{im21}(A)Z_{lm17}(B)]\\
&= \mathsf{E}[Z_{im18}(A)Z_{lm22}(B)]\\
&= -\mathsf{E}[Z_{im22}(A)Z_{lm18}(B)]\\
&= \mathsf{E}[Z_{im19}(A)Z_{lm23}(B)]\\
&= -\mathsf{E}[Z_{im23}(A)Z_{lm19}(B)]\\
&= \mathsf{E}[Z_{im20}(A)Z_{lm24}(B)]\\
&= -\mathsf{E}[Z_{im24}(A)Z_{lm20}(B)]\\
&= \int\limits_{A\cap B} \nu_{E_u,11}(\vec{k}) \left(-\frac{1}{2}\mathsf{f}^{E_u,11}_{il} + \frac{\sqrt{3}}{2}\mathsf{f}^{E_u,12}_{il}\right) \mathrm{d}\mu_m(\vec{k})\,.
\end{aligned}$$

Proof Only one detail requires additional explanation. We choose $I_{E_g} = I_{E_u} = \{1\}$.

Theorem 4.8 *Let* $\mathsf{d}(\vec{x})$ *be a homogeneous and* $(\mathcal{T} \times Z_2^c, A_u \oplus E_u)$*-isotropic random field. We have*

$$\langle \mathsf{d}(\vec{x})\rangle = 0\,.$$

In the basis (3.8), the two-point correlation tensor of the field has the form

$$\begin{aligned}
\langle \mathsf{d}(\vec{x}), \mathsf{d}(\vec{y})\rangle = &\sum_{m=0}^{5} \int\limits_{(\hat{\mathbb{R}}^3/\mathcal{T}\times Z_2^c)_m} j^{A_g}(\vec{k}, \vec{y}-\vec{x})\mathsf{f}^{\mathcal{T}\times Z_2^c,m,A_g}(\vec{k})\,\mathrm{d}\mu_m(\vec{k})\\
&+ \sum_{m\in\{1,3,4,5\}} \int\limits_{(\hat{\mathbb{R}}^3/\mathcal{T}\times Z_2^c)_m} \sum_{k=1}^{2} j^{E_g}_{k1}(\vec{k}, \vec{y}-\vec{x}) \sum_{q=1}^{2} \nu_{E_g,q1}(\vec{k})\mathsf{f}^{E_g,qk}\,\mathrm{d}\mu_m(\vec{k})\,,
\end{aligned}$$

where the spherical Bessel functions $j^{A_g}(\vec{k}, \vec{y} - \vec{x})$ *and* $j^{E_g}(\vec{k}, \vec{y} - \vec{x})$ *are given by Eqs. (4.57) and (4.58), and the basis tensors* $\mathsf{f}^{E_g, ik}$ *are given by Eqs. (4.25) and (4.26). The field has the form*

$$\mathsf{d}(\vec{x}) = \frac{1}{\sqrt{3}} \sum_{i=1}^{3} \sum_{m=0}^{5} \sum_{n=1}^{24} \int_{(\hat{\mathbb{R}}^3/\mathcal{T} \times Z_2^c)_m} j_n(\vec{k}, \vec{x}) \, \mathrm{d}Z_{imn}(\vec{k}) \mathsf{f}^i \,,$$

where $j_n(\vec{k}, \vec{x})$ *are 8 different combinations of cosines and sines of* $k_1 x_1$, $k_2 x_2$, *and* $k_3 x_3$ *for* $1 \leq n \leq 8$, *of cosines and sines of* $k_1 x_2$, $k_2 x_3$, *and* $k_3 x_1$ *for* $9 \leq n \leq 16$, *and of cosines and sines of* $k_1 x_3$, $k_2 x_1$, *and* $k_3 x_2$ *for* $17 \leq n \leq 24$. $Z_{imn}(\vec{k})$ *are centred real-valued random measures on* $(\hat{\mathbb{R}}^3/\mathcal{T} \times Z_2^c)_m$ *with the following nonzero correlations. For* $m \in \{0, 2\}$

$$\mathsf{E}[Z_{imn}(A) Z_{lmn}(B)] = \int_{A \cap B} \mathsf{f}_{il}^{\mathcal{T} \times Z_2^c, m, A_g}(\vec{k}) \, \mathrm{d}\mu_m(\vec{k}) \,.$$

For $m \in \{1, 3, 4, 5\}$ *and* $1 \leq n \leq 8$

$$\mathsf{E}[Z_{imn}(A) Z_{lmn}(B)] = \int_{A \cap B} \left[\mathsf{f}_{il}^{\mathcal{T} \times Z_2^c, m, A_g}(\vec{k}) + \sum_{q=1}^{2} v_{E_g, q1}(\vec{k}) \mathsf{f}_{il}^{E_g, q1} \right] \mathrm{d}\mu_m(\vec{k}) \,.$$

For $m \in \{1, 3, 4, 5\}$ *and* $9 \leq n \leq 16$

$$\begin{aligned}\mathsf{E}[Z_{imn}(A) Z_{lmn}(B)] = \int_{A \cap B} & \left[\mathsf{f}_{il}^{\mathcal{T} \times Z_2^c, m, A_g}(\vec{k}) \right. \\ & \left. + \sum_{q=1}^{2} v_{E_g, q1}(\vec{k}) \left(-\frac{1}{2} \mathsf{f}_{il}^{E_g, q1} - \frac{\sqrt{3}}{2} \mathsf{f}_{il}^{E_g, q2} \right) \right] \mathrm{d}\mu_m(\vec{k}) \,.\end{aligned}$$

For $m \in \{1, 3, 4, 5\}$ *and* $17 \leq n \leq 24$

$$\begin{aligned}\mathsf{E}[Z_{imn}(A) Z_{lmn}(B)] = \int_{A \cap B} & \left[\mathsf{f}_{il}^{\mathcal{T} \times Z_2^c, m, A_g}(\vec{k}) \right. \\ & \left. + \sum_{q=1}^{2} v_{E_g, q1}(\vec{k}) \left(-\frac{1}{2} \mathsf{f}_{il}^{E_g, q1} + \frac{\sqrt{3}}{2} \mathsf{f}_{il}^{E_g, q2} \right) \right] \mathrm{d}\mu_m(\vec{k}) \,.\end{aligned}$$

Theorem 4.9 *Let* $\mathsf{d}(\vec{x})$ *be a homogeneous and* $(O, A_2 \oplus E)$*-isotropic random field. We have*

$$\langle \mathsf{d}(\vec{x}) \rangle = 0 .$$

In the basis (3.8), the two-point correlation tensor of the field has the form

$$\begin{aligned}\langle \mathsf{d}(\vec{x}), \mathsf{d}(\vec{y}) \rangle &= \sum_{m=0}^{7} \int_{(\hat{\mathbb{R}}^3/O\times Z_2^c)_m} j^{A_{1g}}(\vec{k}, \vec{y}-\vec{x}) \mathsf{f}^{O\times Z_2^c,m,A_{1g}}(\vec{k})\,\mathrm{d}\mu_m(\vec{k})\\ &+ \sum_{m\in\{1,3,4,5,6,7\}} \int_{(\hat{\mathbb{R}}^3/O\times Z_2^c)_m} \sum_{k=1}^{2} j_{k1}^{E_g}(\vec{k}, \vec{y}-\vec{x}) \sum_{q=1}^{2} \nu_{E_g,q1}(\vec{k}) \mathsf{f}^{E_g,qk}\,\mathrm{d}\mu_m(\vec{k})\\ &+ \sum_{m\in\{2,4,5,7\}} \int_{(\hat{\mathbb{R}}^3/O\times Z_2^c)_m} j^{A_{2u}}(\vec{k}, \vec{y}-\vec{x}) \mathsf{f}^{O\times Z_2^c,m,A_{2u}}(\vec{k})\,\mathrm{d}\mu_m(\vec{k})\\ &+ \sum_{m\in\{5,7\}} \int_{(\hat{\mathbb{R}}^3/O\times Z_2^c)_m} \nu_{E_u,11}(\vec{k}) \sum_{k=1}^{2} j_{k1}^{E_u}(\vec{k}, \vec{y}-\vec{x}) \mathsf{f}^{E_u,1k}\,\mathrm{d}\mu_m(\vec{k}) ,\end{aligned}$$

where the spherical Bessel functions $j^{A_{1g}}(\vec{k}, \vec{y}-\vec{x})$, …, $j^{E_u}(\vec{k}, \vec{y}-\vec{x})$ *are given by Eqs. (4.60) and (4.61), and the basis tensors* $\mathsf{f}^{E_g,ik}$ *and* $\mathsf{f}^{E_u,ik}$ *are given by Eqs. (4.25), (4.26), and (4.28). The field has the form*

$$\mathsf{d}(\vec{x}) = \frac{1}{\sqrt{6}} \sum_{i=1}^{3} \sum_{m=0}^{7} \sum_{n=1}^{48} \int_{(\hat{\mathbb{R}}^3/O\times Z_2^c)_m} j_n(\vec{k}, \vec{x})\,\mathrm{d}Z_{mn}(\vec{k}) \mathsf{f}^i ,$$

where $j_n(\vec{k}, \vec{x})$ *are 8 different combinations of cosines and sines of* k_1x_1, k_2x_2, *and* k_3x_3 *for* $1 \le n \le 8$, *of* k_1x_1, k_2x_3, *and* k_3x_2 *for* $9 \le n \le 16$, *of* k_1x_3, k_2x_1, *and* k_3x_2 *for* $17 \le n \le 24$, *of* k_1x_2, k_2x_3, *and* k_3x_1 *for* $25 \le n \le 32$, *of* k_1x_3, k_2x_1, *and* k_3x_2 *for* $33 \le n \le 40$, *and of* k_1x_3, k_2x_2, *and* k_3x_1 *for* $41 \le n \le 48$. $Z_{imn}(\vec{k})$ *are centred real-valued random measures on* $(\hat{\mathbb{R}}^3/\mathcal{T} \times Z_2^c)_m$ *with the following nonzero correlations. For* $m \in \{0, 2\}$

$$\mathsf{E}[Z_{imn}(A) Z_{lmn}(B)] = \int_{A\cap B} \mathsf{f}_{il}^{O\times Z_2^c,m,A_g}(\vec{k})\,\mathrm{d}\mu_m(\vec{k}) .$$

For $m \in \{1, 3, 4, 5, 6, 7\}$ *and* $1 \le n \le 8$, $17 \le n \le 24$

$$\mathsf{E}[Z_{imn}(A)Z_{lmn}(B)] = \int_{A\cap B} \left[\mathsf{f}_{il}^{O\times Z_2^c,m,A_g}(\vec{k}) + \sum_{q=1}^{2} \nu_{E_g,q1}(\vec{k})\mathsf{f}_{il}^{E_g,q1} \right] \mathrm{d}\mu_m(\vec{k}).$$

For $m \in \{1, 3, 4, 5\}$ *and* $9 \le n \le 16$

$$\mathsf{E}[Z_{imn}(A)Z_{lmn}(B)] = \int_{A\cap B} \left[\mathsf{f}_{il}^{O\times Z_2^c,m,A_g}(\vec{k}) + \sum_{q=1}^{2} \nu_{E_g,q1}(\vec{k}) \left(\frac{1}{2}\mathsf{f}_{il}^{E_g,q1} - \frac{\sqrt{3}}{2}\mathsf{f}_{il}^{E_g,q2} \right) \right] \mathrm{d}\mu_m(\vec{k}).$$

For $m \in \{1, 3, 4, 5\}$ *and* $25 \le n \le 32$

$$\mathsf{E}[Z_{imn}(A)Z_{lmn}(B)] = \int_{A\cap B} \left[\mathsf{f}_{il}^{O\times Z_2^c,m,A_g}(\vec{k}) + \sum_{q=1}^{2} \nu_{E_g,q1}(\vec{k}) \left(-\frac{1}{2}\mathsf{f}_{il}^{E_g,q1} + \frac{\sqrt{3}}{2}\mathsf{f}_{il}^{E_g,q2} \right) \right] \mathrm{d}\mu_m(\vec{k}).$$

For $m \in \{1, 3, 4, 5\}$ *and* $33 \le n \le 40$

$$\mathsf{E}[Z_{imn}(A)Z_{lmn}(B)] = \int_{A\cap B} \left[\mathsf{f}_{il}^{O\times Z_2^c,m,A_g}(\vec{k}) + \sum_{q=1}^{2} \nu_{E_g,q1}(\vec{k}) \left(-\frac{1}{2}\mathsf{f}_{il}^{E_g,q1} - \frac{\sqrt{3}}{2}\mathsf{f}_{il}^{E_g,q2} \right) \right] \mathrm{d}\mu_m(\vec{k}).$$

For $m \in \{1, 3, 4, 5\}$ *and* $41 \le n \le 48$

$$\mathsf{E}[Z_{imn}(A)Z_{lmn}(B)] = \int_{A\cap B} \left[\mathsf{f}_{il}^{O\times Z_2^c,m,A_g}(\vec{k}) + \sum_{q=1}^{2} \nu_{E_g,q1}(\vec{k}) \left(\frac{1}{2}\mathsf{f}_{il}^{E_g,q1} + \frac{\sqrt{3}}{2}\mathsf{f}_{il}^{E_g,q2} \right) \right] \mathrm{d}\mu_m(\vec{k}).$$

For $m \in \{2,4\}$ *and* $0 \le k \le 5$

$$\begin{aligned}
\mathsf{E}[Z_{im8k+1}(A)Z_{lm8k+8}(B)] &= -\mathsf{E}[Z_{im8k+8}(A)Z_{lm8k+1}(B)] \\
&= -\mathsf{E}[Z_{im8k+2}(A)Z_{lm8k+7}(B)] \\
&= \mathsf{E}[Z_{im8k+7}(A)Z_{lm8k+2}(B)] \\
&= -\mathsf{E}[Z_{im8k+3}(A)Z_{lm8k+6}(B)] \\
&= \mathsf{E}[Z_{im8k+6}(A)Z_{lm8k+3}(B)] \\
&= \mathsf{E}[Z_{im8k+4}(A)Z_{lm8k+5}(B)] \\
&= -\mathsf{E}[Z_{im8k+5}(A)Z_{lm8k+4}(B)] \\
&= \int\limits_{A\cap B} \mathsf{f}_{il}^{O\times Z_2^c,m,A_{2u}}(\vec{k})\,\mathrm{d}\mu_m(\vec{k})\,.
\end{aligned}$$

For $m \in \{5,7\}$ *and* $k \in \{0,2\}$

$$\begin{aligned}
\mathsf{E}[Z_{im8k+1}(A)Z_{lm8k+8}(B)] &= -\mathsf{E}[Z_{im8k+8}(A)Z_{lm8k+1}(B)] \\
&= -\mathsf{E}[Z_{im8k+2}(A)Z_{lm8k+7}(B)] \\
&= \mathsf{E}[Z_{im8k+7}(A)Z_{lm8k+2}(B)] \\
&= -\mathsf{E}[Z_{im8k+3}(A)Z_{lm8k+6}(B)] \\
&= \mathsf{E}[Z_{im8k+6}(A)Z_{lm8k+3}(B)] \\
&= \mathsf{E}[Z_{im8k+4}(A)Z_{lm8k+5}(B)] \\
&= -\mathsf{E}[Z_{im8k+5}(A)Z_{lm8k+4}(B)] \\
&= \int\limits_{A\cap B} \left[\mathsf{f}_{il}^{O\times Z_2^c,m,A_{2u}}(\vec{k}) + \mathrm{i}\nu_{E_u,11}(\vec{k})\mathsf{f}_{il}^{E_u,11}\right]\mathrm{d}\mu_m(\vec{k})\,.
\end{aligned}$$

For $m \in \{5,7\}$ *and* $9 \le n \le 16$

$$\begin{aligned}
&\mathsf{E}[Z_{im9}(A)Z_{lm16}(B)] = -\mathsf{E}[Z_{im16}(A)Z_{lm9}(B)] \\
&= -\mathsf{E}[Z_{im10}(A)Z_{lm15}(B)] = \mathsf{E}[Z_{im15}(A)Z_{lm10}(B)] \\
&= -\mathsf{E}[Z_{im11}(A)Z_{lm14}(B)] = \mathsf{E}[Z_{im14}(A)Z_{lm11}(B)] \\
&= \mathsf{E}[Z_{im12}(A)Z_{lm13}(B)] = -\mathsf{E}[Z_{im13}(A)Z_{lm12}(B)] \\
&= \int\limits_{A\cap B} \left[\mathsf{f}_{il}^{O\times Z_2^c,m,A_{2u}}(\vec{k}) + \mathrm{i}\nu_{E_u,11}(\vec{k})\left(-\frac{1}{2}\mathsf{f}_{il}^{E_u,11} + \frac{\sqrt{3}}{2}\mathsf{f}_{il}^{E_u,12}\right)\right]\mathrm{d}\mu_m(\vec{k})\,.
\end{aligned}$$

For $m \in \{5,7\}$ *and* $25 \le n \le 32$

$$\begin{aligned}
&\mathsf{E}[Z_{im25}(A)Z_{lm32}(B)] = -\mathsf{E}[Z_{im32}(A)Z_{lm25}(B)]\\
&= -\mathsf{E}[Z_{im26}(A)Z_{lm31}(B)] = \mathsf{E}[Z_{im31}(A)Z_{lm26}(B)]\\
&= -\mathsf{E}[Z_{im27}(A)Z_{lm30}(B)] = \mathsf{E}[Z_{im30}(A)Z_{lm27}(B)]\\
&= \mathsf{E}[Z_{im28}(A)Z_{lm29}(B)] = -\mathsf{E}[Z_{im29}(A)Z_{lm28}(B)]\\
&= \int_{A\cap B}\left[\mathsf{f}^{O\times Z^c_2,m,A_{2u}}_{il}(\vec{k}) + \mathrm{i}\nu_{E_u,11}(\vec{k})\left(\frac{1}{2}\mathsf{f}^{E_u,11}_{il} - \frac{\sqrt{3}}{2}\mathsf{f}^{E_u,12}_{il}\right)\right]\,\mathrm{d}\mu_m(\vec{k})\,.
\end{aligned}$$

For $m \in \{5,7\}$ *and* $33 \le n \le 40$

$$\begin{aligned}
&\mathsf{E}[Z_{im33}(A)Z_{lm40}(B)] = -\mathsf{E}[Z_{im40}(A)Z_{lm33}(B)]\\
&= -\mathsf{E}[Z_{im34}(A)Z_{lm39}(B)] = \mathsf{E}[Z_{im39}(A)Z_{lm34}(B)]\\
&= -\mathsf{E}[Z_{im35}(A)Z_{lm38}(B)] = \mathsf{E}[Z_{im38}(A)Z_{lm35}(B)]\\
&= \mathsf{E}[Z_{im36}(A)Z_{lm37}(B)] = -\mathsf{E}[Z_{im37}(A)Z_{lm36}(B)]\\
&= \int_{A\cap B}\left[\mathsf{f}^{O\times Z^c_2,m,A_{2u}}_{il}(\vec{k}) + \mathrm{i}\nu_{E_u,11}(\vec{k})\left(\frac{1}{2}\mathsf{f}^{E_u,11}_{il} + \frac{\sqrt{3}}{2}\mathsf{f}^{E_u,12}_{il}\right)\right]\,\mathrm{d}\mu_m(\vec{k})\,.
\end{aligned}$$

For $m \in \{5,7\}$ *and* $41 \le n \le 48$

$$\begin{aligned}
&\mathsf{E}[Z_{im41}(A)Z_{lm48}(B)] = -\mathsf{E}[Z_{im48}(A)Z_{lm41}(B)]\\
&= -\mathsf{E}[Z_{im42}(A)Z_{lm47}(B)] = \mathsf{E}[Z_{im47}(A)Z_{lm42}(B)]\\
&= -\mathsf{E}[Z_{im43}(A)Z_{lm46}(B)] = \mathsf{E}[Z_{im46}(A)Z_{lm43}(B)]\\
&= \mathsf{E}[Z_{im44}(A)Z_{lm45}(B)] = -\mathsf{E}[Z_{im45}(A)Z_{lm44}(B)]\\
&= \int_{A\cap B}\left[\mathsf{f}^{O\times Z^c_2,m,A_{2u}}_{il}(\vec{k}) + \mathrm{i}\nu_{E_u,11}(\vec{k})\left(-\frac{1}{2}\mathsf{f}^{E_u,11}_{il} - \frac{\sqrt{3}}{2}\mathsf{f}^{E_u,12}_{il}\right)\right]\,\mathrm{d}\mu_m(\vec{k})\,.
\end{aligned}$$

Theorem 4.10 *Let* $\mathsf{d}(\vec{x})$ *be a homogeneous and* $(O^-, A_1 \oplus E)$*-isotropic random field. We have*

$$\langle \mathsf{d}(\vec{x})\rangle = C_1\mathsf{d}^1\,,$$

where d^1 *is the first vector of the basis (3.8). In the above basis, the two-point correlation tensor of the field has the form*

$$
\begin{aligned}
\langle \mathsf{d}(\vec{x}), \mathsf{d}(\vec{y}) \rangle &= \sum_{m=0}^{7} \int\limits_{(\hat{\mathbb{R}}^3/O\times Z_2^c)_m} j^{A_{1g}}(\vec{k}, \vec{y}-\vec{x}) \mathsf{f}^{O^-,m,A_{1g}}(\vec{k})\, \mathrm{d}\mu_m(\vec{k}) \\
&+ \sum_{m\in\{1,3,4,5,6,7\}} \int\limits_{(\hat{\mathbb{R}}^3/O\times Z_2^c)_m} \sum_{k=1}^{2} j^{E_g}_{k1}(\vec{k}, \vec{y}-\vec{x}) \sum_{q=1}^{2} v_{E_g,q1}(\vec{k}) \mathsf{f}^{E_g,qk}\, \mathrm{d}\mu_m(\vec{k}) \\
&+ \int\limits_{(\hat{\mathbb{R}}^3/O\times Z_2^c)_7} j^{A_{1u}}(\vec{k}, \vec{y}-\vec{x}) \mathsf{f}^{O^-,m,A_{1u}}(\vec{k})\, \mathrm{d}\mu_m(\vec{k}) \\
&+ \sum_{m\in\{5,7\}} \int\limits_{(\hat{\mathbb{R}}^3/O\times Z_2^c)_m} v_{E_u,11}(\vec{k}) \sum_{k=1}^{2} j^{E_u}_{k1}(\vec{k}, \vec{y}-\vec{x}) \mathsf{f}^{E_u,1k}\, \mathrm{d}\mu_m(\vec{k}),
\end{aligned}
$$

where the spherical Bessel functions $j^{A_{1g}}(\vec{k}, \vec{y}-\vec{x})$, …, $j^{E_u}(\vec{k}, \vec{y}-\vec{x})$ *are given by Eqs. (4.60), (4.61), and (4.62), and the basis tensors* $\mathsf{f}^{E_g,ik}$ *and* $\mathsf{f}^{E_u,ik}$ *are given by Eqs. (4.25), (4.26), and (4.28). The field has the form*

$$
\mathsf{d}(\vec{x}) = C_1 \mathsf{d}^1 + \frac{1}{\sqrt{6}} \sum_{i=1}^{3} \sum_{m=0}^{7} \sum_{n=1}^{48} \int\limits_{(\hat{\mathbb{R}}^3/O\times Z_2^c)_m} j_n(\vec{k}, \vec{x})\, \mathrm{d}Z_{imn}(\vec{k}) \mathsf{f}^i ,
$$

where $j_n(\vec{k}, \vec{x})$ *are 8 different combinations of cosines and sines of* k_1x_1, k_2x_2, *and* k_3x_3 *for* $1 \le n \le 8$, *of* k_1x_1, k_2x_3, *and* k_3x_2 *for* $9 \le n \le 16$, *of* k_1x_3, k_2x_1, *and* k_3x_2 *for* $17 \le n \le 24$, *of* k_1x_2, k_2x_3, *and* k_3x_1 *for* $25 \le n \le 32$, *of* k_1x_3, k_2x_1, *and* k_3x_2 *for* $33 \le n \le 40$, *and of* k_1x_3, k_2x_2, *and* k_3x_1 *for* $41 \le n \le 48$. $Z_{imn}(\vec{k})$ *are centred real-valued random measures on* $(\hat{\mathbb{R}}^3/\mathcal{T} \times Z_2^c)_m$ *with the following nonzero correlations. For* $m \in \{0, 2\}$

$$
\mathsf{E}[Z_{imn}(A) Z_{lmn}(B)] = \int\limits_{A\cap B} \mathsf{f}_{il}^{O^-,m,A_g}(\vec{k})\, \mathrm{d}\mu_m(\vec{k}) .
$$

For $m \in \{1, 3, 4, 5, 6, 7\}$ *and* $1 \le n \le 8$, $17 \le n \le 24$

$$
\mathsf{E}[Z_{imn}(A) Z_{lmn}(B)] = \int\limits_{A\cap B} \left[\mathsf{f}_{il}^{O^-,m,A_g}(\vec{k}) + \sum_{q=1}^{2} v_{E_g,q1}(\vec{k}) \mathsf{f}_{il}^{E_g,q1} \right] \mathrm{d}\mu_m(\vec{k}) .
$$

For $m \in \{1, 3, 4, 5\}$ *and* $9 \le n \le 16$

$$
\begin{aligned}
\mathsf{E}[Z_{imn}(A) Z_{lmn}(B)] = \int\limits_{A\cap B} \Bigg[&\mathsf{f}_{il}^{O^-,m,A_g}(\vec{k}) \\
&+ \sum_{q=1}^{2} v_{E_g,q1}(\vec{k}) \left(\frac{1}{2} \mathsf{f}_{il}^{E_g,q1} - \frac{\sqrt{3}}{2} \mathsf{f}_{il}^{E_g,q2} \right) \Bigg] \mathrm{d}\mu_m(\vec{k}) .
\end{aligned}
$$

For $m \in \{1, 3, 4, 5\}$ *and* $25 \le n \le 32$

$$\mathsf{E}[Z_{imn}(A)Z_{lmn}(B)] = \int\limits_{A\cap B} \Bigg[\mathsf{f}_{il}^{O^-,m,A_g}(\vec{k}) + \sum_{q=1}^{2} v_{E_g,q1}(\vec{k}) \left(-\frac{1}{2}\mathsf{f}_{il}^{E_g,q1} + \frac{\sqrt{3}}{2}\mathsf{f}_{il}^{E_g,q2}\right)\Bigg] \,\mathrm{d}\mu_m(\vec{k})\,.$$

For $m \in \{1, 3, 4, 5\}$ *and* $33 \le n \le 40$

$$\mathsf{E}[Z_{imn}(A)Z_{lmn}(B)] = \int\limits_{A\cap B} \Bigg[\mathsf{f}_{il}^{O^-,m,A_g}(\vec{k}) + \sum_{q=1}^{2} v_{E_g,q1}(\vec{k}) \left(-\frac{1}{2}\mathsf{f}_{il}^{E_g,q1} - \frac{\sqrt{3}}{2}\mathsf{f}_{il}^{E_g,q2}\right)\Bigg] \,\mathrm{d}\mu_m(\vec{k})\,.$$

For $m \in \{1, 3, 4, 5\}$ *and* $41 \le n \le 48$

$$\mathsf{E}[Z_{imn}(A)Z_{lmn}(B)] = \int\limits_{A\cap B} \Bigg[\mathsf{f}_{il}^{O^-,m,A_g}(\vec{k}) + \sum_{q=1}^{2} v_{E_g,q1}(\vec{k}) \left(\frac{1}{2}\mathsf{f}_{il}^{E_g,q1} + \frac{\sqrt{3}}{2}\mathsf{f}_{il}^{E_g,q2}\right)\Bigg] \,\mathrm{d}\mu_m(\vec{k})\,.$$

For $m = 7$ *and* $k \in \{0, 2\}$

$$\begin{aligned}
\mathsf{E}[Z_{i7\,8k+1}(A)Z_{l7\,8k+8}(B)] &= -\mathsf{E}[Z_{i7\,8k+8}(A)Z_{l7\,8k+1}(B)] \\
&= -\mathsf{E}[Z_{i7\,8k+2}(A)Z_{l7\,8k+7}(B)] \\
&= \mathsf{E}[Z_{i7\,8k+7}(A)Z_{l7\,8k+2}(B)] \\
&= -\mathsf{E}[Z_{i7\,8k+3}(A)Z_{l7\,8k+6}(B)] \\
&= \mathsf{E}[Z_{i7\,8k+6}(A)Z_{l7\,8k+3}(B)] \\
&= \mathsf{E}[Z_{i7\,8k+4}(A)Z_{l7\,8k+5}(B)] \\
&= -\mathsf{E}[Z_{i7\,8k+5}(A)Z_{l7\,8k+4}(B)] \\
&= \int\limits_{A\cap B} \left[\mathsf{f}_{il}^{O^-,7,A_{1u}}(\vec{k}) + \mathrm{i}v_{E_u,11}(\vec{k})\mathsf{f}_{il}^{E_u,11}\right] \mathrm{d}\mu_7(\vec{k})\,.
\end{aligned}$$

For $m = 7$ and $9 \leq n \leq 16$

$$\begin{aligned}
&\mathsf{E}[Z_{i7\,9}(A)Z_{l7\,16}(B)] = -\mathsf{E}[Z_{i7\,16}(A)Z_{l7\,9}(B)]\\
&= -\mathsf{E}[Z_{i7\,10}(A)Z_{l7\,15}(B)] = \mathsf{E}[Z_{i7\,15}(A)Z_{l7\,10}(B)]\\
&= -\mathsf{E}[Z_{i7\,11}(A)Z_{l7\,14}(B)] = \mathsf{E}[Z_{i7\,14}(A)Z_{l7\,11}(B)]\\
&= \mathsf{E}[Z_{i7\,12}(A)Z_{l7\,13}(B)] = -\mathsf{E}[Z_{i7\,13}(A)Z_{l7\,12}(B)]\\
&= \int\limits_{A\cap B} \left[-\mathsf{f}_{il}^{O^-,7,A_{1u}}(\vec{k}) + \mathrm{i}\nu_{E_u,11}(\vec{k})\left(-\frac{1}{2}\mathsf{f}_{il}^{E_u,11} + \frac{\sqrt{3}}{2}\mathsf{f}_{il}^{E_u,12}\right)\right] \mathrm{d}\mu_7(\vec{k})\,.
\end{aligned}$$

For $m = 7$ and $25 \leq n \leq 32$

$$\begin{aligned}
&\mathsf{E}[Z_{i7\,25}(A)Z_{l7\,32}(B)] = -\mathsf{E}[Z_{i7\,32}(A)Z_{l7\,25}(B)]\\
&= -\mathsf{E}[Z_{i7\,26}(A)Z_{l7\,31}(B)] = \mathsf{E}[Z_{i7\,31}(A)Z_{l7\,26}(B)]\\
&= -\mathsf{E}[Z_{i7\,27}(A)Z_{l7\,30}(B)] = \mathsf{E}[Z_{i7\,30}(A)Z_{l7\,27}(B)]\\
&= \mathsf{E}[Z_{i7\,28}(A)Z_{l7\,29}(B)] = -\mathsf{E}[Z_{i7\,29}(A)Z_{l7\,28}(B)]\\
&= \int\limits_{A\cap B} \left[-\mathsf{f}_{il}^{O^-,7,A_{1u}}(\vec{k}) + \mathrm{i}\nu_{E_u,11}(\vec{k})\left(\frac{1}{2}\mathsf{f}_{il}^{E_u,11} - \frac{\sqrt{3}}{2}\mathsf{f}_{il}^{E_u,12}\right)\right] \mathrm{d}\mu_7(\vec{k})\,.
\end{aligned}$$

For $m = 7$ and $33 \leq n \leq 40$

$$\begin{aligned}
&\mathsf{E}[Z_{i7\,33}(A)Z_{l7\,40}(B)] = -\mathsf{E}[Z_{i7\,40}(A)Z_{l7\,33}(B)]\\
&= -\mathsf{E}[Z_{i7\,34}(A)Z_{l7\,39}(B)] = \mathsf{E}[Z_{i7\,39}(A)Z_{l7\,34}(B)]\\
&= -\mathsf{E}[Z_{i7\,35}(A)Z_{l7\,38}(B)] = \mathsf{E}[Z_{i7\,38}(A)Z_{l7\,35}(B)]\\
&= \mathsf{E}[Z_{i7\,36}(A)Z_{l7\,37}(B)] = -\mathsf{E}[Z_{i7\,37}(A)Z_{l7\,36}(B)]\\
&= \int\limits_{A\cap B} \left[-\mathsf{f}_{il}^{O^-,7,A_{1u}}(\vec{k}) + \mathrm{i}\nu_{E_u,11}(\vec{k})\left(\frac{1}{2}\mathsf{f}_{il}^{E_u,11} + \frac{\sqrt{3}}{2}\mathsf{f}_{il}^{E_u,12}\right)\right] \mathrm{d}\mu_7(\vec{k})\,.
\end{aligned}$$

For $m = 7$ and $41 \leq n \leq 48$

$$\begin{aligned}
&\mathsf{E}[Z_{i7\,41}(A)Z_{l7\,48}(B)] = -\mathsf{E}[Z_{i7\,48}(A)Z_{l7\,41}(B)]\\
&= -\mathsf{E}[Z_{i7\,42}(A)Z_{l7\,47}(B)] = \mathsf{E}[Z_{i7\,47}(A)Z_{l7\,42}(B)]\\
&= -\mathsf{E}[Z_{i7\,43}(A)Z_{l7\,46}(B)] = \mathsf{E}[Z_{i7\,46}(A)Z_{l7\,43}(B)]\\
&= \mathsf{E}[Z_{i7\,44}(A)Z_{l7\,45}(B)] = -\mathsf{E}[Z_{i7\,45}(A)Z_{l7\,44}(B)]\\
&= \int\limits_{A\cap B} \left[\mathsf{f}_{il}^{O^-,7,A_{1u}}(\vec{k}) + \mathrm{i}\nu_{E_u,11}(\vec{k})\left(-\frac{1}{2}\mathsf{f}_{il}^{E_u,11} - \frac{\sqrt{3}}{2}\mathsf{f}_{il}^{E_u,12}\right)\right] \mathrm{d}\mu_7(\vec{k})\,.
\end{aligned}$$

For $m = 5$ and $k \in \{0, 2\}$

$$
\begin{aligned}
\mathsf{E}[Z_{i5\,8k+1}(A)Z_{l5\,8k+8}(B)] &= -\mathsf{E}[Z_{i5\,8k+8}(A)Z_{l5\,8k+1}(B)] \\
&= -\mathsf{E}[Z_{i5\,8k+2}(A)Z_{l5\,8k+7}(B)] \\
&= \mathsf{E}[Z_{i5\,8k+7}(A)Z_{l5\,8k+2}(B)] \\
&= -\mathsf{E}[Z_{i5\,8k+3}(A)Z_{l5\,8k+6}(B)] \\
&= \mathsf{E}[Z_{i5\,8k+6}(A)Z_{l5\,8k+3}(B)] \\
&= \mathsf{E}[Z_{i5\,8k+4}(A)Z_{l5\,8k+5}(B)] \\
&= -\mathsf{E}[Z_{i5\,8k+5}(A)Z_{l5\,8k+4}(B)] \\
&= \int\limits_{A\cap B} \mathrm{i}\nu_{E_u,11}(\vec{k})\mathsf{f}_{il}^{E_u,11}\,\mathrm{d}\mu_5(\vec{k}).
\end{aligned}
$$

For $m = 5$ and $9 \le n \le 16$

$$
\begin{aligned}
&\mathsf{E}[Z_{i5\,9}(A)Z_{l5\,16}(B)] = -\mathsf{E}[Z_{i5\,16}(A)Z_{l5\,9}(B)] \\
&= -\mathsf{E}[Z_{i5\,10}(A)Z_{l5\,15}(B)] = \mathsf{E}[Z_{i5\,15}(A)Z_{l5\,10}(B)] \\
&= -\mathsf{E}[Z_{i5\,11}(A)Z_{l5\,14}(B)] = \mathsf{E}[Z_{i5\,14}(A)Z_{l5\,11}(B)] \\
&= \mathsf{E}[Z_{i5\,12}(A)Z_{l5\,13}(B)] = -\mathsf{E}[Z_{i5\,13}(A)Z_{l5\,12}(B)] \\
&= \int\limits_{A\cap B} \left[\mathrm{i}\nu_{E_u,11}(\vec{k})\left(-\frac{1}{2}\mathsf{f}_{il}^{E_u,11} + \frac{\sqrt{3}}{2}\mathsf{f}_{il}^{E_u,12}\right)\right]\mathrm{d}\mu_5(\vec{k}).
\end{aligned}
$$

For $m = 5$ and $25 \le n \le 32$

$$
\begin{aligned}
&\mathsf{E}[Z_{i5\,25}(A)Z_{l5\,32}(B)] = -\mathsf{E}[Z_{i5\,32}(A)Z_{l5\,25}(B)] \\
&= -\mathsf{E}[Z_{i5\,26}(A)Z_{l5\,31}(B)] = \mathsf{E}[Z_{i5\,31}(A)Z_{l5\,26}(B)] \\
&= -\mathsf{E}[Z_{i5\,27}(A)Z_{l5\,30}(B)] = \mathsf{E}[Z_{i5\,30}(A)Z_{l5\,27}(B)] \\
&= \mathsf{E}[Z_{i5\,28}(A)Z_{l5\,29}(B)] = -\mathsf{E}[Z_{i5\,29}(A)Z_{l5\,28}(B)] \\
&= \int\limits_{A\cap B} \left[\mathrm{i}\nu_{E_u,11}(\vec{k})\left(\frac{1}{2}\mathsf{f}_{il}^{E_u,11} - \frac{\sqrt{3}}{2}\mathsf{f}_{il}^{E_u,12}\right)\right]\mathrm{d}\mu_5(\vec{k}).
\end{aligned}
$$

For $m = 5$ and $33 \le n \le 40$

$$
\begin{aligned}
&\mathsf{E}[Z_{i5\,33}(A)Z_{l5\,40}(B)] = -\mathsf{E}[Z_{i5\,40}(A)Z_{l5\,33}(B)] \\
&= -\mathsf{E}[Z_{i5\,34}(A)Z_{l5\,39}(B)] = \mathsf{E}[Z_{i5\,39}(A)Z_{l5\,34}(B)] \\
&= -\mathsf{E}[Z_{i5\,35}(A)Z_{l5\,38}(B)] = \mathsf{E}[Z_{i5\,38}(A)Z_{l5\,35}(B)] \\
&= \mathsf{E}[Z_{i5\,36}(A)Z_{l5\,37}(B)] = -\mathsf{E}[Z_{i5\,37}(A)Z_{l5\,36}(B)] \\
&= \int\limits_{A\cap B} \left[\mathrm{i}\nu_{E_u,11}(\vec{k})\left(\frac{1}{2}\mathsf{f}_{il}^{E_u,11} + \frac{\sqrt{3}}{2}\mathsf{f}_{il}^{E_u,12}\right)\right]\mathrm{d}\mu_5(\vec{k}).
\end{aligned}
$$

For $m = 5$ *and* $41 \le n \le 48$

$$\begin{aligned}
&\mathsf{E}[Z_{i5\,41}(A)Z_{l5\,48}(B)] = -\mathsf{E}[Z_{i5\,48}(A)Z_{l5\,41}(B)]\\
&= -\mathsf{E}[Z_{i5\,42}(A)Z_{l5\,47}(B)] = \mathsf{E}[Z_{i5\,47}(A)Z_{l5\,42}(B)]\\
&= -\mathsf{E}[Z_{i5\,43}(A)Z_{l5\,46}(B)] = \mathsf{E}[Z_{i5\,46}(A)Z_{l5\,43}(B)]\\
&= \mathsf{E}[Z_{i5\,44}(A)Z_{l5\,45}(B)] = -\mathsf{E}[Z_{i5\,45}(A)Z_{l5\,44}(B)]\\
&= \int_{A\cap B}\left[\mathrm{i}\nu_{E_u,11}(\vec{k})\left(-\frac{1}{2}\mathsf{f}^{E_u,11}_{il}-\frac{\sqrt{3}}{2}\mathsf{f}^{E_u,12}_{il}\right)\right]\,\mathrm{d}\mu_5(\vec{k})\,.
\end{aligned}$$

Theorem 4.11 *Let* $\mathsf{d}(\vec{x})$ *be a homogeneous and* $(O \times Z^c_2, A_{2u} \oplus E_u)$*-isotropic random field. We have*

$$\langle \mathsf{d}(\vec{x})\rangle = 0\,.$$

In the basis (3.8), the two-point correlation tensor of the field has the form

$$\begin{aligned}
\langle \mathsf{d}(\vec{x}), \mathsf{d}(\vec{y})\rangle &= \sum_{m=0}^{7}\int_{(\hat{\mathbb{R}}^3/O\times Z^c_2)_m} j^{A_{1g}}(\vec{k}, \vec{y}-\vec{x})\mathsf{f}^{O\times Z^c_2,m,A_{1g}}(\vec{k})\,\mathrm{d}\mu_m(\vec{k})\\
&+ \sum_{m\in\{1,3,4,5,6,7\}}\int_{(\hat{\mathbb{R}}^3/O\times Z^c_2)_m}\sum_{k=1}^{2} j^{E_g}_{k1}(\vec{k}, \vec{y}-\vec{x})\sum_{q=1}^{2}\nu_{E_g,q1}(\vec{k})\mathsf{f}^{E_g,qk}\,\mathrm{d}\mu_m(\vec{k})\,,
\end{aligned}$$

where the spherical Bessel functions $j^{A_{1g}}(\vec{k}, \vec{y}-\vec{x})$ *and* $j^{E_g}(\vec{k}, \vec{y}-\vec{x})$ *are given by Eqs. (4.60) and (4.61), and the basis tensors* $\mathsf{f}^{E_g,ik}$ *and* $\mathsf{f}^{E_u,ik}$ *are given by Eqs. (4.25), (4.26), and (4.28). The field has the form*

$$\mathsf{d}(\vec{x}) = \frac{1}{\sqrt{6}}\sum_{i=1}^{3}\sum_{m=0}^{7}\sum_{n=1}^{48}\int_{(\hat{\mathbb{R}}^3/O\times Z^c_2)_m} j_n(\vec{k}, \vec{x})\,\mathrm{d}Z_{imn}(\vec{k})\mathsf{f}^i\,,$$

where $j_n(\vec{k}, \vec{x})$ *are 8 different combinations of cosines and sines of* k_1x_1, k_2x_2, *and* k_3x_3 *for* $1 \le n \le 8$, *of* k_1x_1, k_2x_3, *and* k_3x_2 *for* $9 \le n \le 16$, *of* k_1x_3, k_2x_1, *and* k_3x_2 *for* $17 \le n \le 24$, *of* k_1x_2, k_2x_3, *and* k_3x_1 *for* $25 \le n \le 32$, *of* k_1x_3, k_2x_1, *and* k_3x_2 *for* $33 \le n \le 40$, *and of* k_1x_3, k_2x_2, *and* k_3x_1 *for* $41 \le n \le 48$. $Z_{imn}(\vec{k})$ *are centred real-valued random measures on* $(\hat{\mathbb{R}}^3/\mathcal{T} \times Z^c_2)_m$ *with the following nonzero correlations. For* $m \in \{0, 2\}$

$$\mathsf{E}[Z_{imn}(A)Z_{lmn}(B)] = \int_{A\cap B}\mathsf{f}^{O\times Z^c_2,m,A_g}_{il}(\vec{k})\,\mathrm{d}\mu_m(\vec{k})\,.$$

For $m \in \{1, 3, 4, 5, 6, 7\}$ *and* $1 \leq n \leq 8$, $17 \leq n \leq 24$

$$\mathsf{E}[Z_{imn}(A)Z_{lmn}(B)] = \int_{A\cap B} \left[\mathsf{f}_{il}^{O\times Z_2^c,m,A_g}(\vec{k}) + \sum_{q=1}^{2} \nu_{E_g,q1}(\vec{k})\mathsf{f}^{E_g,q1} \right] \mathrm{d}\mu_m(\vec{k}).$$

For $m \in \{1, 3, 4, 5\}$ *and* $9 \leq n \leq 16$

$$\begin{aligned}\mathsf{E}[Z_{imn}(A)Z_{lmn}(B)] = \int_{A\cap B} & \left[\mathsf{f}_{il}^{O\times Z_2^c,m,A_g}(\vec{k}) \right.\\ & \left. + \sum_{q=1}^{2} \nu_{E_g,q1}(\vec{k}) \left(\frac{1}{2}\mathsf{f}_{il}^{E_g,q1} - \frac{\sqrt{3}}{2}\mathsf{f}_{il}^{E_g,q2} \right) \right] \mathrm{d}\mu_m(\vec{k}).\end{aligned}$$

For $m \in \{1, 3, 4, 5\}$ *and* $25 \leq n \leq 32$

$$\begin{aligned}\mathsf{E}[Z_{imn}(A)Z_{lmn}(B)] = \int_{A\cap B} & \left[\mathsf{f}_{il}^{O\times Z_2^c,m,A_g}(\vec{k}) \right.\\ & \left. + \sum_{q=1}^{2} \nu_{E_g,q1}(\vec{k}) \left(-\frac{1}{2}\mathsf{f}_{il}^{E_g,q1} + \frac{\sqrt{3}}{2}\mathsf{f}_{il}^{E_g,q2} \right) \right] \mathrm{d}\mu_m(\vec{k}).\end{aligned}$$

For $m \in \{1, 3, 4, 5\}$ *and* $33 \leq n \leq 40$

$$\begin{aligned}\mathsf{E}[Z_{imn}(A)Z_{lmn}(B)] = \int_{A\cap B} & \left[\mathsf{f}_{il}^{O\times Z_2^c,m,A_g}(\vec{k}) \right.\\ & \left. + \sum_{q=1}^{2} \nu_{E_g,q1}(\vec{k}) \left(-\frac{1}{2}\mathsf{f}_{il}^{E_g,q1} - \frac{\sqrt{3}}{2}\mathsf{f}_{il}^{E_g,q2} \right) \right] \mathrm{d}\mu_m(\vec{k}).\end{aligned}$$

For $m \in \{1, 3, 4, 5\}$ *and* $41 \leq n \leq 48$

$$\begin{aligned}\mathsf{E}[Z_{imn}(A)Z_{lmn}(B)] = \int_{A\cap B} & \left[\mathsf{f}_{il}^{O\times Z_2^c,m,A_g}(\vec{k}) \right.\\ & \left. + \sum_{q=1}^{2} \nu_{E_g,q1}(\vec{k}) \left(\frac{1}{2}\mathsf{f}_{il}^{E_g,q1} + \frac{\sqrt{3}}{2}\mathsf{f}_{il}^{E_g,q2} \right) \right] \mathrm{d}\mu_m(\vec{k}).\end{aligned}$$

Example 4.7 Let $\mathsf{d}(\vec{x})$ be a homogeneous and $(O \times Z_2^c, A_{2u} \oplus E_u)$-isotropic random field. Assume that the spectral measure of the *homogeneous* random field $\mathsf{d}(\vec{x})$ is absolutely continuous with respect to the Lebesgue measure on the wavenumber domain $\hat{\mathbb{R}}^3$. In that case, the measures μ_m, $1 \leq m \leq 7$, are zero measures, because they are supported by the sets $(\hat{\mathbb{R}}^3/O \times Z_2^c)_m$ whose dimensions are less than 3. The closure of the set $(\hat{\mathbb{R}}^3/O \times Z_2^c)_0$ is the set $\Lambda = \{\vec{k} \in \hat{\mathbb{R}}^3 : 0 \leq k_1 \leq k_2 \leq k_3\}$. The measure μ_0 has the form

$$\mu_0(A) = \int_A u(\vec{k})\,\mathrm{d}\vec{k},$$

where $u(\vec{k})\colon \Lambda \to [0,\infty)$ is a measurable function with

$$\int_\Lambda u(\vec{k})\,\mathrm{d}\vec{k} < \infty,$$

and where A is an arbitrary Borel subset of Λ. Theorem 4.11 gives $\langle \mathsf{d}(\vec{x})\rangle = \mathsf{0}$. Equation (4.60) gives

$$\begin{aligned}\langle \mathsf{d}(\vec{x}), \mathsf{d}(\vec{y})\rangle = \frac{1}{6}\int_\Lambda &\{\cos(k_1z_1)[\cos(k_2z_2)\cos(k_3z_3) + \cos(k_2z_3)\cos(k_3z_2)]\\ &+ \cos(k_2z_1)[\cos(k_1z_3)\cos(k_3z_2) + \cos(k_1z_2)\cos(k_3z_3)]\\ &+ \cos(k_3z_1)[\cos(k_1z_2)\cos(k_2z_3) + \cos(k_1z_3)\cos(k_2z_2)]\}\\ &\times \mathsf{f}^{O\times Z_2^c,0,A_{1g}}(\vec{k})u(\vec{k})\,\mathrm{d}\vec{k},\end{aligned}$$

where $\mathsf{f}^{O\times Z_2^c,0,A_{1g}}(\vec{k})$ is a measurable function on Λ that takes values in the set (4.38). This set is the closed interval with extreme points $\mathsf{f}^{A_g,11}$ and $\frac{1}{\sqrt{2}}\mathsf{f}^{A_g,21}$, where the above matrices are given by (4.24). We obtain

$$\mathsf{f}^{O\times Z_2^c,0,A_{1g}}(\vec{k}) = \lambda(\vec{k})\mathsf{f}^{A_g,11} + \frac{1-\lambda(\vec{k})}{\sqrt{2}}\mathsf{f}^{A_g,21},$$

where $\lambda(\vec{k})\colon \Lambda \to [0,1]$ is an arbitrary measurable function. The two-point correlation tensor of the random field $\mathsf{d}(\vec{x})$ in the basis (3.8) takes the form

$$\begin{aligned}\langle \mathsf{d}(\vec{x}), \mathsf{d}(\vec{y})\rangle = \int_\Lambda &\{\cos(k_1z_1)[\cos(k_2z_2)\cos(k_3z_3) + \cos(k_2z_3)\cos(k_3z_2)]\\ &+ \cos(k_2z_1)[\cos(k_1z_3)\cos(k_3z_2) + \cos(k_1z_2)\cos(k_3z_3)]\\ &+ \cos(k_3z_1)[\cos(k_1z_2)\cos(k_2z_3) + \cos(k_1z_3)\cos(k_2z_2)]\}\\ &\times \begin{pmatrix} 1&0&0\\ 0&0&0\\ 0&0&0\end{pmatrix} u_1(\vec{k})\,\mathrm{d}\vec{k}\\ &+ \int_\Lambda \{\cos(k_1z_1)[\cos(k_2z_2)\cos(k_3z_3) + \cos(k_2z_3)\cos(k_3z_2)]\\ &+ \cos(k_2z_1)[\cos(k_1z_3)\cos(k_3z_2) + \cos(k_1z_2)\cos(k_3z_3)]\\ &+ \cos(k_3z_1)[\cos(k_1z_2)\cos(k_2z_3) + \cos(k_1z_3)\cos(k_2z_2)]\}\\ &\times \begin{pmatrix} 0&0&0\\ 0&1&0\\ 0&0&1\end{pmatrix} u_2(\vec{k})\,\mathrm{d}\vec{k},\end{aligned}$$

where

$$u_1(\vec{k}) = \frac{\lambda(\vec{k})}{6}u(\vec{k}), \qquad u_2(\vec{k}) = \frac{1-\lambda(\vec{k})}{12\sqrt{2}}u(\vec{k}).$$

The random field $\mathsf{d}(\vec{x})$ has the form

$$\mathsf{d}(\vec{x}) = \sum_{i=1}^{3} \sum_{n=1}^{48} \int_{\Lambda} j_n(\vec{k}, \vec{x})\, \mathrm{d}Z_n^{(i)}(\vec{k}) \mathsf{f}^i ,$$

where the functions $j_n(\vec{k}, \vec{x})$ are described in Theorem 4.11, the tensors f^i are given by (3.8), and $Z_n^{(i)}$ are centred real-valued random measures on Λ with nonzero correlations

$$\mathsf{E}[Z_n^{(1)}(A) Z_n^{(1)}(B)] = \int_{A \cap B} u_1(\vec{k})\, \mathrm{d}\vec{k} ,$$

$$\mathsf{E}[Z_n^{(2)}(A) Z_n^{(2)}(B)] = \int_{A \cap B} u_2(\vec{k})\, \mathrm{d}\vec{k} ,$$

$$\mathsf{E}[Z_n^{(3)}(A) Z_n^{(3)}(B)] = \int_{A \cap B} u_2(\vec{k})\, \mathrm{d}\vec{k} .$$

References

1. Adams, J.F.: Lectures on Lie Groups. W. A. Benjamin Inc, New York-Amsterdam (1969)
2. Altmann, S.L., Herzig, P.: Point-Group Theory Tables. Oxford Science Publications. Clarendon Press (1994). https://books.google.se/books?id=2R_wAAAAMAAJ
3. Amiri-Hezaveh, A., Karimi, P., Ostoja-Starzewski, M.: Stress field formulation of linear electro-magneto-elastic materials. Math. Mech. Solids **24**(12), 3806–3822 (2019). https://doi.org/10.1177/1081286519857127
4. Amiri-Hezaveh, A., Karimi, P., Ostoja-Starzewski, M.: IBVP for electromagneto-elastic materials: variational approach. Math. Mech. Complex Syst. **8**(1), 47–67 (2020). https://doi.org/10.2140/memocs.2020.8.47
5. Berezans'kiĭ, Yu.M.: Expansions in eigen functions of selfadjoint operators. In: Bolstein, R., Danskin, J.M., Rovnyak, J., Shulman L. (eds.) Translations of Mathematical Monographs, vol. 17. American Mathematical Society, Providence, R.I. (1968). Translated from the Russian
6. Bourbaki, N.: Integration. II. Chapters 7–9. Elements of Mathematics (Berlin). Springer-Verlag, Berlin (2004). Translated from the 1963 and 1969 French originals by Sterling K. Berberian
7. Cramér, H.: On the theory of stationary random processes. Ann. Math. **2**(41), 215–230 (1940). https://doi.org/10.2307/1968827
8. Eringen, A., Maugin, G.A.: Electrodynamics of Continua I. Springer, Foundations and Solid Media. Springer (1990)
9. Godunov, S.K., Gordienko, V.M.: Clebsch-Gordan coefficients in the case of various choices of bases of unitary and orthogonal representations of the groups SU(2) and SO(3). Sibirsk. Mat. Zh. **45**(3), 540–557 (2004). https://doi.org/10.1023/B:SIMJ.0000028609.97557.b8
10. Golubitsky, M., Stewart, I., Schaeffer, D.G.: Singularities and groups in bifurcation theory, Vol. II, Applied Mathematical Sciences, vol. 69. Springer-Verlag, New York (1988). https://doi.org/10.1007/978-1-4612-4574-2
11. Gurtin, M.E.: Variational principles in the linear theory of viscoelasticity. Arch. Rational Mech. Anal. **13**, 179–191 (1963). https://doi.org/10.1007/BF01262691
12. Gurtin, M.E.: Variational principles for linear elastodynamics. Arch. Rational Mech. Anal. **16**, 34–50 (1964). https://doi.org/10.1007/BF00248489
13. Hofmann, K.H., Morris, S.A.: The structure of compact groups. A primer for the student—a handbook for the expert, De Gruyter Studies in Mathematics, vol. 25, 3rd edn. De Gruyter, Berlin (2013)
14. Ignaczak, J.: Direct determination of stresses from the stress equations of motion in elasticity. Arch. Mech. Stos. **11**, 671–678 (1959)

A. Malyarenko et al., *Random Fields of Piezoelectricity and Piezomagnetism*, SpringerBriefs in Mathematical Methods,
https://doi.org/10.1007/978-3-030-60064-8

15. Ignaczak, J.: A completeness problem for stress equations of motion in the linear elasticity theory. Arch. Mech. Stos. **15**, 225–234 (1963)
16. Ihrig, E., Golubitsky, M.: Pattern selection with O(3) symmetry. Phys. D **13**(1–2), 1–33 (1984). https://doi.org/10.1016/0167-2789(84)90268-9
17. Karimi, P., Zhang, X., Yan, S., Ostoja-Starzewski, M., Jin, J.M.: Electrostatic and magnetostatic properties of random materials. Phys. Rev. E **99**, 022,120 (2019). https://doi.org/10.1103/PhysRevE.99.022120. https://link.aps.org/doi/10.1103/PhysRevE.99.022120
18. Khisaeva, Z.F., Ostoja-Starzewski, M.: On the size of RVE in finite elasticity of random composites. J. Elasticity **85**(2), 153–173 (2006). https://doi.org/10.1007/s10659-006-9076-y
19. Li, J.Y.: Uniqueness and reciprocity theorems for linear thermo-electro-magneto-elasticity. Quart. J. Mech. Appl. Math. **56**(1), 35–43 (2003). https://doi.org/10.1093/qjmam/56.1.35
20. Malyarenko, A., Ostoja-Starzewski, M.: Tensor-Valued Random Fields for Continuum Physics. Cambridge Monographs on Mathematical Physics. Cambridge University Press, Cambridge (2019)
21. Maugin, G.A.: Continuum Mechanics of Electromagnetic Solids, North-Holland Series in Applied Mathematics and Mechanics, vol. 33. Wiley, Amsterdam (1988). https://doi.org/10.1002/zamm.19890691106
22. Miehe, C., Rosato, D., Kiefer, B.: Variational principles in dissipative electro-magneto-mechanics: a framework for the macro-modeling of functional materials. Int. J. Numer. Meth. Eng. **86**(10), 1225–1276 (2011)
23. Olive, M.: Effective computation of SO(3) and O(3) linear representation symmetry classes. Math. Mech. Compl. Sys. **7**(3), 203–237 (2019). https://doi.org/10.102140/memocs.2019.7.203
24. Olive, M., Auffray, N.: Symmetry classes for odd-order tensors. ZAMM Z. Angew. Math. Mech. **94**(5), 421–447 (2014). https://doi.org/10.1002/zamm.201200225
25. Ostoja-Starzewski, M.: Ignaczak equation of elastodynamics. Math. Mech. Solids **24**(11), 3674–3713 (2019). https://doi.org/10.1177/1081286518757284
26. Ostoja-Starzewski, M., Kale, S., Karimi, P., Malyarenko, A., Raghavan, B., Ranganathan, S., Zhang, J.: Chapter Two—scaling to RVE in random media. pp. 111–211. Elsevier (2016). https://doi.org/10.1016/bs.aams.2016.07.001. http://www.sciencedirect.com/science/article/pii/S0065215616300011
27. Pérez-Fernández, L.D., Bravo-Castillero, J., Rodríguez-Ramos, R., Sabina, F.J.: On the constitutive relations and energy potentials of linear thermo-magneto-electro-elasticity. Mech. Res. Comm. **36**(3), 343–350 (2009). https://doi.org/10.1016/j.mechrescom.2008.10.003
28. Qin, Q.H.: Advanced Mechanics of Piezoelectricity. Higher Education Press, Beijing; Springer, Heidelberg (2013). https://doi.org/10.1007/978-3-642-29767-0
29. Ranganathan, S.I., Ostoja-Starzewski, M.: Scaling function, anisotropy and the size of RVE in elastic random polycrystals. J. Mech. Phys. Solids **56**(9), 2773–2791 (2008). https://doi.org/10.1016/j.jmps.2008.05.001
30. Ranganathan, S.I., Ostoja-Starzewski, M.: Towards scaling laws in random polycrystals. Internat. J. Engrg. Sci. **47**(11–12), 1322–1330 (2009). https://doi.org/10.1016/j.ijengsci.2009.05.003
31. Selivanova, S.: Computing Clebsch-Gordan matrices with applications in elasticity theory. In: Logic, Computation, Hierarchies, Ontos Math. Log., vol. 4, pp. 273–295. De Gruyter, Berlin (2014)
32. Sternberg, S.: Group Theory and Physics. Cambridge University Press, Cambridge (1994)

Index

A. Malyarenko et al., *Random Fields of Piezoelectricity and Piezomagnetism*,
SpringerBriefs in Mathematical Methods,
https://doi.org/10.1007/978-3-030-60064-8

Printed in the United States
By Bookmasters